Conversations About Neuroscience

Conversations About

NEUROSCIENCE

Edited by Howard Burton

R Ideas Roadshow
INTELLIGENT. INQUISITIVE. INTERNATIONAL.

Ideas Roadshow conversations present a wealth of candid insights from some of the world's leading experts, generated through a focused yet informal setting. They are explicitly designed to give non-specialists a uniquely accessible window into frontline research and scholarship that wouldn't otherwise be encountered through standard lectures and textbooks.

Over 100 Ideas Roadshow conversations have been held since our debut in 2012, covering a wide array of topics across the arts and sciences.

All Ideas Roadshow conversations are available both as part of a collection or as an individual eBook.

See www.ideasroadshow.com for a full listing of all titles.

Edited, with preface and all introductions written by Howard Burton.

All *Ideas Roadshow Conversations* use Canadian spelling.

Contents

VISION AND PERCEPTION
A CONVERSATION WITH KALANIT GRILL-SPECTOR

INVESTIGATING INTELLIGENCE
A CONVERSATION WITH JOHN DUNCAN

MINDS AND MACHINES
A CONVERSATION WITH MIGUEL NICOLELIS

Textual Note

The contents of this book are based upon separate filmed conversations with Howard Burton and each of the five featured experts.

Lisa Feldman Barrett is University Distinguished Professor in Psychology at Northeastern University. This conversation occurred on August 30, 2015.

Jennifer Groh is Professor of Psychology and Neuroscience, Neurobiology and Computer Science at Duke University. This conversation occurred on October 30, 2014.

Kalanit Grill-Spector is Professor in Psychology and the Stanford Neurosciences Institute at Stanford University. This conversation occurred on March 20, 2014.

John Duncan is Professor of Cognitive Neuroscience and Programme leader, Executive processes group of the MRC Cognition and Brain Sciences Unit at the University of Cambridge. This conversation occurred on August 31, 2012.

Miguel Nicolelis is Professor of Neurobiology, Neurology, Neurosurgery, Biomedical Engineering, Psychology and Neuroscience and Orthopaedic Surgery and Co-Director of the Center for Neuroengineering at Duke University. This conversation occurred on April 22, 2013.

Howard Burton is the creator and host of Ideas Roadshow and was Founding Executive Director of Perimeter Institute for Theoretical Physics.

Preface

One of the intriguing things that happens when you have the occasion to talk to a wide variety of different experts in a particular field is not only the connections that link similar ideas across separate conversations, but the *differences* in approach that also become visible.

When it comes to neuroscience, two such differences come immediately to the fore, each of which has its own distinct historical context, and each of which is on display in this collection.

The first is the issue of localization vs. distribution.

Best exemplified by the 18th-century German physician and founder of phrenology, Frans Joseph Gall, the localized perspective attributes specific functionality to corresponding physical regions of the brain. Although most now believe that Gall held an extreme sort of position—with phrenology these days often invoked as a paradigmatic example of a "pseudoscience"—the view that the brain is composed of separate processing regions for different tasks still lingers, not only in the public consciousness in terms of a simplified "left-brain, right-brain" distinction of cognitive function (or, rather worse still, "left brain" or "right brain" personality characterizations), but also in terms of particular subdisciplines of neuroscience where such localized perspectives have proven to be particularly effective.

Perhaps the most famous of these subdisciplines is vision, where a well-established framework of visual processing has developed over many decades, culminating in a theory of two processing streams that traverse a number of specifically defined brain regions—labelled, rather unimaginatively, V1, V2 and so on. Such a framework is naturally attractive to the likes of vision scientists like **Kalanit**

Grill-Spector who find themselves prone to invoking images of a sequential, localized approach to information processing, much like a computer.

> *"For me vision is a very concrete system: the goal is to see. Then within seeing there are concrete goals that you might want to figure out. For example, recognizing people—that's the 'what' stream or ventral stream; figuring out where things are—that's the dorsal stream.*

> *"These processing streams are composed of components called areas, and what we have to figure out is how many areas you have in the stream and then how information gets relayed from one area to another area."*

For **John Duncan**, however, decades of research on intelligence led him to the conclusion that a key brain feature associated with John Spearman's celebrated "g factor" isn't a highly-localized brain region at all, but rather an extended network that is distributed over a large area of the brain, touching on both the frontal and parietal lobes.

> *"We can ask, 'What is it in the brain that is active while people solve puzzles of this sort?' And it turns out that the answer is somewhat complicated but also surprisingly simple: it's far from the whole brain that is involved when you're doing tasks of this sort. Instead, it's a very particular network in the frontal lobes—right behind your forehead—and the parietal lobe, which is sort of on top and to the rear, if you like. Very interestingly, if you gave many different sorts of tasks—tests of memory or language, or even identifying faces or holding something in short term memory—the same network tends to be a part of the brain's response. And much of what we're doing now is trying to understand what's going on inside these regions of the brain as problems are solved."*

Miguel Nicolelis, meanwhile, extends the distributed network thesis even further, insisting that the central feature of **all** brain processing is such distributed networks, notwithstanding the brain-lesioning arguments that others invoke to demonstrate the priority of localized processing regions.

"You need a grid for everything to work properly. It's pretty much like the internet. If you take a few peripheral servers out, nothing happens. But if you were to take Google out, all the servers that Google has, the internet will feel the effect. Because now you have a massive disconnection of a key hub.

"So in my view none of that is incompatible with the distributed hypothesis. Classically, these examples have been used to suggest a localization of function: there is a degree of specialization, no doubt about it. But it's not as strict as we were led to believe, and it's not phrenology, not by a long shot."

Beyond this localized vs. distributed question lies a second core theme, the extent to which the brain is an active or passive participant in interpreting the world around us.

Kalanit Grill-Spector, as you might expect, focuses on how the brain processes visual incoming signals from the outside world.

"Basically our goal is to understand how information gets from the eye, goes through the optic nerve to the occipital lobe, and then through a mysterious sequence of processing you get this "Aha! moment". This is what we are trying to figure out, this sequence of processing different brain function."

For **Jennifer Groh**, though, the brain is much more than a passive recipient of external stimuli—a key concern for her is understanding the mechanisms of how, precisely, the brain actively coordinates the vast array of information offered to us through our different senses.

"The auditory system is computing sound location based on cues that are fundamentally anchored to the head, while the visual system is computing visual locations based on cues that are fundamentally anchored to the orientation of the eyes. So every time your eyes move, you're yanking your visual scene around to some new position with respect to your auditory scene.

"I've been really interested in how the brain fixes that, how it puts those two signals into a common frame of reference so that you can do things like use lip-reading cues to help you understand what someone is saying. If you don't correctly associate the lips of the person who's speaking with the sound of that person speaking, then you can't make use of that supplementary information."

Lisa Feldman Barrett, on the other hand, doesn't just opt to focus on select active aspects of the brain: she is deeply convinced (along with **Miguel Nicolelis**) that the active nature of our brain is nothing less than its fundamental defining quality.

"Your brain is not wired to be dormant until it's stimulated by something in the environment, which it then processes and reacts to. In fact, what's happening if you look at the anatomy of the brain, if you take a physiological approach in trying to understand metabolic efficiency, you see that what the brain is doing is predicting what sensory information will be coming and comparing those predictions to the information that actually arrives: it's forming perceptions of emotional experience, of somebody else's emotion—of everything, really—in this predictive way; and it's using prior experience, which it organizes as concepts."

There is, unsurprisingly perhaps, a strong correspondence between views on these two issues. Those who primarily regard the brain as a processing centre of incoming information tend to focus on localized areas where such processing occurs, while those who ascribe a much more active role for the brain in regularly predicting its surrounding environment tend to give precedence to regionally distributed networks for brain processes.

So, who is right?

Well, that's not, I believe, the right question. Both have a litany of solid experimental justification for their views—obviously nobody is suggesting, for instance, that over 60 years of increasingly accurate models of the brain's visual system be suddenly disregarded.

Moreover, it's clear that *both* are necessary to developing a coherent overall framework of the situation at hand: no matter how convinced you might be that the brain is actively making predictions about its surrounding environment, say, it is obviously essential to fully understand how, exactly, it processes the consequent information to determine if its predictions were actually correct.

But what *is* at stake, it seems, is the development of underlying, guiding principles—not just understanding the mechanics of specific processes going on in our heads, in short, but **why**. To take one very important example, understanding what, exactly, neuroplasticity is, why it exists and how it might be successfully harnessed, will necessarily differ significantly depending on one's perspective of how the brain is operating at a very fundamental level.

Life at the front lines of scientific research is always messy and cacophonous in real time, suffused with a lively exchange of differing opinions and perspectives—and it's hard to imagine more of a cutting-edge scientific venture these days than neuroscience.

Pull up a chair and watch the show.

Constructing Our World

The Brain's-Eye View

A conversation with Lisa Feldman Barrett

Introduction

Putting the Pieces Together

One of the fascinating aspects of having the opportunity to talk to lots of different people about lots of different things is that you start seeing interesting connections when you least expect it.

I knew that Lisa Feldman Barrett was a highly accomplished scientist before I sat down to talk to her, but I hadn't fully appreciated how reflective and wide-ranging she was, and I certainly hadn't expected that we'd spend as much time talking about structural and philosophical approaches to scientific inquiry as we did. Of course, we met a couple of years before her bestselling popular book, *How Emotions Are Made: The Secret Life of the Brain*, came out, so all I had to go on were strong recommendations from colleagues and various academic papers.

I'd imagined that we would focus on the specifics of her scientific work on emotion, such as her pioneering "Conceptual-Act Model", and that's the sort of thing I had read about going into the discussion. But very quickly things took a most interesting turn.

"Ernst Mayr made a very important point, which psychologists haven't quite embraced yet, that Darwin's greatest contribution is not the concept of natural selection: it's the idea that a species is a conceptual category.

"It's not a physical category with necessary and sufficient features, and firm boundaries, and a biological essence. In fact, that's sort of the "dog-show version" of evolution: like there's a perfect cocker spaniel and all the other variations of cocker spaniels are some kind of error, because there really is a pure, Platonic form of cocker

spanielness that you see in award-winning cocker spaniels who have exactly the right nose size, and exactly the right coat thickness, and so on.

"One of Darwin's points was that a species is a conceptual category because the instances—of cocker spaniels, let's say—vary from one another, and that variation is meaningful—it's meaningfully tied to the situation, to the environment.

"The idea that variability is meaningful, and that really all biological categories are actually concepts—because, within a category, things can vary quite a bit, and the name of the category refers to a collection of things, a population of things—is called "population thinking". This is why natural selection works—precisely because such variability actually exists and is important."

Suddenly, I started thinking of an entirely different conversation, one with award-winning geneticist Stephen Scherer. In 2004, Stephen was having a devil of a time getting his groundbreaking article on his groundbreaking discovery that so-called large-scale copy number variation to our genomes was vastly more common than previously believed. While people had long known that there were certain conditions resulting from situations where huge strings of our DNA were repeated or deleted, such as Down's Syndrome, the common consensus of the international genetic community in the early 2000s was that the *only* genetic variation that occurred for the rest of us were mutations of individual nucleotides.

Well, you might think to yourself, that's strange. After all, didn't the Human Genome Project—arguably one of the most comprehensive international scientific collaborations in human history finish up by around that time?

Indeed it did. And that was exactly the problem. Because the way the Human Genome Project was structured was to look at over 700 different donors and rigorously compare their DNA, site by site, along the entire genome to produce a "consensus sequence". And anything which didn't fit the consensus would be discarded. Which means that they naturally missed all the large-scale variation—in other words, it was simply treated as error.

Sounds familiar?

Lisa is not a geneticist, of course–although there's doubtless still some room in her winding career path from pre-med student to clinical psychologist to social psychologist to cognitive neuroscientist to one day incorporate that too—but the central point remains: ignoring the inherent value in variability is a natural consequence of what Lisa views as the strong spirit of "essentialism" that has long pervaded psychology.

> *"There has always been a debate about whether or not there are essential characteristics—either essential features or an essential element—that define each psychology category as a single thing.*

> *"But it's also true that alongside that, there have always been philosophers and scientists who have argued against that kind of essentialism. If you look at the history of philosophy of mind, the two perspectives have really just proceeded in parallel with no resolution, and usually essentialism dominates.*

> *"In psychology, we would call that "faculty" psychology: the idea that the mind is best described as a set of characteristics or abilities and each one is elemental and has some natural essence or a natural aspect to it that's basic and elemental.*

> *"So when I say something like, "There's no physical signature for anger," people assume I'm saying, "Anger is not real." I'm not saying that anger is not real. I'm saying that anger is a conceptual category: instances of anger are highly variable, and we have to try to understand that variability and how it is that all instances of anger can all be anger while still having that variability.*

But for Lisa, conceptual categories aren't simply aspects of a key philosophical argument, they are the building blocks to her entire approach to the brain—including, but hardly limited to emotion:

"What's happening is that, every waking moment of your life, your brain is anticipating and making sense of sensory inputs from its environment—the combination of the internal environment of the body and the external environment—and it's using conceptual knowledge to do that. That's basically it.

"The theory is not domain specific. I'm not saying this only works for emotion. I'm saying this is really how the brain works: it works this way when you're constructing a visual perception, when you're constructing an auditory perception, when you're constructing an emotion, when you're thinking, and so on.

"In some ways it's a very parsimonious theory, because it doesn't invent any new mechanisms or processes. It's just saying that this is how it works, and this is how it works for virtually all experience."

But however determined she is, Lisa is the furthest thing from dogmatic. In fact, at one point—somewhere between references to neural network homeostasis and Buddhist dharmas—she explicitly welcomed the opportunity for a productive interchange with her "essentialist" opponents:

"I think one thing that really needs to happen is that there has to be a real dialogue about this, not where different sides are caricaturing each other, but a dialogue where they are actually learning from each other what the key experiments need to be in order to move forward."

Appropriately enough for a pioneering constructive neuroscientist, the road to progress is all about making connections.

The Conversation

I. Beginnings

A winding road

HB: I thought we'd talk about your beginnings and your interest in what was, I'm supposing, then psychology—not so much cognitive science—what peaked your interest there and how you got going.

LFB: Well, when I was in high school, I was primarily focused on the sciences and I took some history, but we didn't have psychology in high schools at that time.

HB: This was in Toronto, right?

LFB: Yes, this was in Toronto. So when I went to do my undergraduate degree at the University of Toronto, I thought I was going to be a physician. I thought I was going to medical school, so I took all the pre-med courses, and then I had one more course I had to pick up, so I decided to take a psychology course.

I really took the course because the professor—whose name now escapes me—was really wonderful at teaching and he used to hold his lectures in this place called Convocation Hall, which is a huge lecture theatre, so he would have a thousand students that he would lecture to. My friends loved this course and loved him as a lecturer, so I thought I would just take it and see what I thought.

He was a fabulous lecturer; you felt like he was talking to you specifically, even though you were in this cavernous space. But I really liked the material too. I was just really captivated, by all of it, actually.

HB: What was it, specifically, about that whole experience that captivated you? Do you remember? Was it more him personally or the material?

LFB: He certainly was engaging and he made it very personal. Whereas introductory psychology textbooks are very well known for containing a lot of details that you memorize and not much in the way of connecting those details to everyday life, he was really good at doing that.

And I guess I've just always wondered about people's motivations. I've always had a sense for as long as I can remember that different people can see the same event in very different ways, and here was a course where I was actually given some formal tools to think about these issues in a little bit more precise way.

So I really loved it, and I actually decided to switch my major from biology to psychology. I had to make a decision as to whether or not I was going to major in psychology or anthropology, because I was also taking anthropology courses, which I also loved. In particular, I was really interested in human evolution, so I sort of suffered through cultural anthropology classes, even though now I very much enjoy that perspective too.

I was really fascinated by physical anthropology, so I just took a lot of classes in physiology—both in anthropology and in biology— and I took a lot of psychology classes and some linguistics classes—I think I was one course short of a minor in linguistics.

HB: So you were, in fact, really delving into what is now cognitive science, before its time, as it were.

LFB: Yes, I had no idea. Most of these courses I had never had any access to as a high school student, and I loved my undergraduate experience. I just loved going to classes. I was working two jobs at the same time to pay for it. I had a small scholarship, but I don't come from an affluent background. I had to put myself through school, so I was working two jobs and taking these classes. Even though it was hectic, it was really fun.

HB: Did many other of your fellow psychology students share your perspective on things? I ask because a common stereotype in my experience is, *Most people who take psychology as undergrads are strongly motivated to do so in order to deal with their own issues.*

I'm not necessarily implying that that's true, I'm just saying that that's the stereotype in my experience. What are your thoughts about that? Did you find yourself surrounded by those sorts of people? Is there any merit to that generalization, because you clearly didn't come at it from that sort of perspective at all?

LFB: I guess I would say a couple of things about that. One is, I used to think that psychology—clinical psychology in particular, since that's what I did my PhD in originally—was filled with people who were really nuts—to use a very precise scientific term—and that they were just trying to figure out themselves, "me-search", as some people call it.

But I've discovered that, actually, in academia in general—or maybe just scientists in general—there's just a large proportion of really odd people who think about things in a really odd way, and that's why they are good scientists: because you have to be able to see things other people don't see, you have to notice phenomena other people don't notice, you have to be able to look at things in an unconventional way. I think the proportion is actually high in all the sciences.

I would say—even in biology, and chemistry, and physics—the people I know who went into those fields were doing it very much for personal reasons, they're trying to understand themselves and their place in the universe and that's what takes them into cosmology, say. I still think the proportion is actually really high for all of the scientific disciplines. It's not specific to psychology.

HB: It's just easier to pinpoint with psychology because of the nature of the field.

LFB: Yes, I think so. That's my sense. We're all a little bit nuts, maybe, and that's what maybe helps us be scientists.

HB: OK, so you're captivated by this, you finish your undergraduate degree, and then you think, *I'm going to do a PhD in this sort of thing*. Is that right? Or is that not how it happened?

LFB: It didn't really quite happen like that. I was really on the fence about medicine or a PhD in psychology, and I really couldn't decide up until the last minute. In my fourth year, I applied to only two graduate schools, because I still wasn't sure where I was going to go to graduate school.

I was engaged to be married—my high school sweetheart, actually—and his entire family were all physicians. And everybody I knew, all of my friends, were all going to medical school, or law school, or dentistry—all these professional programs. So I really wasn't sure what I was going to do and there was some pressure to go to medical school.

HB: It seems like there might have been *a lot* of pressure.

LFB: There was a lot of pressure to go to medical school, but I actually loved doing research. And at that point, I had been working in a lab for a couple of years—I worked in a developmental lab and a social psychology lab—so I applied to do a PhD in clinical psychology at the University of Waterloo, which was the best program in the country at the time. It was very well respected in North America. It was seen as quite a terrific program.

Then I applied to do a PhD with Dan Schacter, in cognitive psychology. He was at the University of Toronto and was one of my professors when I was an undergraduate. So I applied to do a degree with him, and I was accepted to both programs. Then Dan decided to take a job at the University of Arizona and I wasn't really in a position where I could go with him, so I decided to do the degree in clinical psychology.

HB: Did that meet your expectations immediately? Did it take a little bit of settling in? How did that work?

LFB: It took a tremendous amount of settling in. In retrospect, I'm glad that I have a clinical degree, but it was not a good fit for me from the beginning, because—well, this is a whole other story, but in clinical psychology, at the time, there was a lot of debate about what the field was. Was it a scientific discipline? Was it a professional field?

These were the kinds of debates that happened in medicine a century ago, but they were happening in clinical psychology at the time when I was actually a graduate student. The University of Waterloo had a fabulous psychology department with many sub-disciplines. It was unusual in that way, that it had really strong people in many sub-disciplines: social psychology, cognitive psychology, and in clinical psychology.

But I was being trained by clinicians in my clinical courses, and I was learning very different things than I was learning in my scientific courses. So I'd take a course in cognitive psychology, say, which was all about heuristics and biases; and then, in my clinical training, I was being trained to do the very things that I knew from my scientific training were the exact opposite of what you would do.

HB: Did you point this out to people?

LFB: I did. It didn't make me a very popular person.

I think that was one issue: that there was a disconnect, because the clinical psychology faculty weren't doing the clinical training—they had brought clinicians in to do that. They were excellent clinicians, mind you, but still there was a real disconnect between the practice and the science, and I think that occurs less now that clinical science is really much more established as a field, but at the time, there was still a lot of debate about it. This was a program that really defined itself as the scientist-practitioner model, and there was this disconnect. So there was that going on.

Then, also, in my scientific endeavours, I was having a pretty difficult time. I was running experiments and they weren't replicating published work; and by the end of my third year, I had run eight experiments, which were all replications—because what you do, as a graduate student, is first run a study that replicates what someone else has done, and then you build on it.

So I would try to run a study and it would fail. Then I would make some tweaks and it would fail again. Then I would try to replicate a different study and go about it a different way to test the hypothesis, and then *it* would fail. Eight times in a row, it failed.

HB: That's statistically significant. Maybe there's something about you.

LFB: Yes, well, that was exactly my thought. I thought, *Well, okay, maybe I'm not cut out to be a scientist.* I was doing pretty well in my clinical training, so I thought, *Well, maybe I'll be a clinician.*

That's a little bit revisionist history, actually—I went into it thinking I'd be a clinician. Even though I loved doing research, being a clinician is still close enough to being a physician that it would still make everybody happy.

I was also fairly insecure as a student, so even though I did very well in university, it didn't occur to me that I was doing very well. I always felt like I wasn't smart enough.

After I had failed to replicate eight experiments, my adviser left to go to Simon Fraser University, so I got a new adviser, Mike Ross, who had just come back from sabbatical. He was in social psychology.

The way I remember it is that he offered to be my adviser and the way that he remembers it is that I asked. But either way, he became my adviser along with another clinical professor, Erik Woody. I made my dissertation proposal and they failed it because it was on this same series of experiments that weren't working.

Mike said, "*The reason I'm failing it is that this is not going to make a good job talk,*" and I thought, *A **job** talk? So he thinks I could actually **do** this as a profession.*

That was a very defining moment for me. I was doing all this other research on the side, but I went back and looked at my eight failed experiments and noticed a pattern in them, and I followed that up, and that became my dissertation. But it was a very windy path for me.

HB: I'd like to move on to the research itself, of course, but first, just a small digression. It certainly strikes me as possible that—arduous though it may have been at the time—your experiences might have helped make you a better supervisor than you might have ordinarily been, because you have a greater degree of empathy and understanding of the anxiety and insecurity that some of your students might have. Do you have that sense as well?

LFB: Absolutely; and we discuss it. I think many graduate students have, at some point in their career—sometimes all the way through, actually—real questions about whether or not they can be scientists and whether or not they're going to be successful.

Graduate school is a really formative time for people, because you start to look at the world in a completely different way than you have before—which can be exhilarating, but also incredibly frightening and distressing. It feels like there's a lot riding on things and everyone's expecting you to be brilliant, expecting you to very quickly produce some brilliant piece of work.

Five years to do a PhD sounds like a long time, but it's actually not a long time. Science doesn't work on the five-year business model. I think that your own sense of insecurity and feelings of distress are not diagnostic of your abilities at all when you are in graduate school.

This is something that I talk about in my lab with individual students. I believe that, as an advisor, my job is not just to teach them how to be scientists in the laboratory, but really how to have a scientific life, a life in science, living the life of the mind.

It requires the ability to doubt yourself—it's good to doubt yourself, to an extent, but also to keep going and not let the doubt actually impair your ability to function and not prevent you from following up on ideas that you are less than 100% sure about. I've never been 100% sure about anything I've ever done.

In fact, I'm usually not even 50% sure. As a scientist, I'm very critically-minded of everybody's work, including my own. So I treat most things with an "as if" quality for many years, because it takes a long time to know whether or not you're on the right track.

HB: I want to talk about emotions and how you got into the study of emotions in particular, but before I do, my understanding is that you also did some sort of fellowship at the University of Manitoba Medical School, right?

LFB: That was my clinical internship. To achieve a PhD in clinical psychology, you must do a year-long internship.

The University of Manitoba was, at the time, the highest ranked internship in Canada. I had looked at other internships, but I decided that Manitoba was the place I wanted to go. They had an excellent program, an excellent reputation, and it also left me additional time to do research. So, while I was on internship, I was also doing research at the University of Manitoba. Many internships wouldn't allow you that flexibility. I was doing my 40 hours a week in my clinical duties and then, in addition, I was doing this work at the University of Manitoba, doing research. They were very accommodating.

HB: I have two questions about that, both of which relate to a conversation I had with Pat Churchland (*Philosophy of Brain*). Before she went to San Diego she was a faculty member in philosophy at Manitoba interested in the philosophy of mind, and people at the University of Manitoba Medical School were quite willing to give her the opportunity to take courses there, which she really wanted to do. It seems they're very open-minded people, those folks in Manitoba.

Anyway, she told me some disgusting anecdote—at least, I thought it was disgusting—about how this experience enabled her to hold a brain in her hands, and how one day she smuggled a brain home to show her philosopher husband and her kids.

I don't remember the timing of things off the top of my head, but my first question is: did you cross paths with her when she was there?

LFB: No.

HB: OK. And my next question is, *Did you also hold a brain in your hand?* Is this a common thing that people in Manitoba do?

LFB: Well, I've never been in that situation, to actually be able hold a brain in my hand. I was more on the teaching end of things than I was taking courses, so I taught medical students while I was there.

HB: Right, but no one threw a brain into your hand when you were teaching medical students?

LFB: No, but I would have loved it if they had.

HB: Uh-huh. OK, back to your story. At this point, were you still vacillating between doing clinical work, or had you been emboldened by the support that had been shown towards you and you were well on your way towards a professional, scientific, academic career?

LFB: Yes, I was pretty much on my way to an academic career, although, at the time, I still thought that I would remain in clinical psychology.

My first job was actually as an assistant professor in a clinical psychology program. Then, after about four years, I formally switched to social personality psychology.

From there, it was just a fairly slow path to neuroscience. First cognitive science and then neuroscience. But it was a long path with a lot of continuous training while I was running a lab.

Questions for Discussion:

1. To what extent do you think that Lisa's complicated, winding career path is indicative of a research area in rapid development? How do you think Lisa's career development would have progressed if she had started her undergraduate experience in today's environment?

2. Do you agree with Lisa's assessment that most people who study science do so for "personal reasons" and are all "a little bit nuts"? Is this any different for those who study non-scientific areas?

II. Confronting Variability

Essentialism vs. conceptual categories

HB: So let's move now to more specifically address your work in emotion, the development of a theory of emotions. How did that begin?

LFB: Well, remember I said that I had these eight experiments that failed, but they all had something in common? What they had in common was that I was measuring feelings of anxiety and depression—not clinical symptoms, but just more like emotions or feeling-states—and the hypothesis had to do with when people feel "anxious" versus when they feel "depressed".

I wasn't ever able to replicate published findings; and when I sat down and looked at the data carefully, across all the studies, I realized that when people reported feeling anxious, they also reported feeling depressed, or they reported feeling neither.

Given the reliability of the measures—how much information was being carried in the measurements—there was no unique variance left over: if people reported feeling one, they reported feeling both, and there was no uniqueness to their reports.

I found this *really* fascinating and wondered why that was the case. In clinical circles, it's still the case that there is tremendous overlap in diagnostic categories—the symptoms overlap, there's co-morbidity in certain anxiety disorders and in major depressive disorders that can be very high for some anxiety disorders, as high as 70% or 80%.

Still, I thought, *Well, do people not **know** the difference between anxiety and depression? Are they **forgetting**? Why is this happening? Does this also happen with other emotions, like anger, and disgust,* *and guilt, and shame? Do people have difficulty distinguishing these too?* It turned out that there are tremendous individual differences in people's tendency to report distinct, emotional experiences.

I thought that I was just going to do this little side project to try to figure out what was going wrong with my measurements and then go back to the original hypothesis that I was really interested in.

HB: What was the original hypothesis?

LFB: It was about self-esteem. It was something called "self-discrepancy theory", which is a theory that Tory Higgins, who's at Columbia University, proposed: the idea that, when you compare your current view of how you are to *your* ideals for yourself, you'll feel depressed if you fall short, while if you compare your current self-assessment to *someone else's* standards for you, you'd feel anxious.

HB: I see. So this is where the anxiousness and the depression aspects come in, which in turn led you to focus on understanding the difficulties in distinguishing between the two.

LFB: Yes. *Was this just these particular subjects? Was it a problem with the measures themselves? What if I used other measures? Was it the fact that, for some reason, these subjects didn't know the difference between anxiety and depression?*

I thought that maybe there was something about the way that I was measuring it that was not forcing people to pay attention to the distinctiveness in those feelings. I was curious about a lot of things.

Eventually, I realized that, in fact, this was not just true of *these* emotions, it's actually true of a much wider set of phenomena.

There is a set of emotions that scientists believed to be basic—meaning inborn, shared with other animals—circuitry in the brain that's specific to those emotions, distinctive displays for those emotions, expressive displays for those emotions, and so forth.

And I was finding in my studies that there were many people who were not distinguishing between what psychologists believed to be very distinctive states, so I was really curious about this. It seemed to me that there was a pretty straightforward way to figure out who was accurate in reporting and who wasn't.

My idea at the time was that some people were very granular in the way that they reported their experiences—they made very precise distinctions. And then there were other people who were using words like "angry", "sad", and "afraid" interchangeably, to mean, "*I feel bad. I'm feeling crappy*"; or "happy", "enthusiastic", and "calm" to mean, "*I feel good. I feel pleasant.*"

So I thought, *These people who are less granular are just not accurately reading off their experiences in their reports. So I just need to teach them how to do that: I just need to teach them to pay attention to the cues and then they'll be more accurate in their reporting.*

Accuracy assumes that there's an actual objective criterion, that there's a perceiver-independent criterion that you can use to say, "*Yes, the anger really **is** present in this person, at this moment*" and that you can compare someone's report to that.

I wasn't trained in emotion theory and neither were my advisers—nobody, actually, at the University of Waterloo had any experience in the science of emotion, but they were very encouraging for whatever I wanted to do.

When I got my first job, I started to seriously investigate this idea, and I systematically went through all these different criteria that were supposed to exist.

Different emotions are supposed to have different facial expressions, so I decided to measure people's faces and measure their facial muscle movements—not just ask one person to perceive emotion in somebody's face, but actually measure specific facial muscles changes. And when you do that, you don't see evidence that different emotions have distinctive, quantitative displays on the face.

So then I thought, *Okay, maybe it's not in the face, but in the body.* Cardiovascular responses are supposed to be distinct for each emotion. William James—the founding father of American psychology—famously said that each emotion has its own physical state.

Well, he *did* say that, but it turns out that he didn't actually mean "each emotion" as in "each emotion category"—he meant "each instance of emotion." So he was talking about instances of anger, say, that can vary tremendously from one another. But that's not how he is cited, that's not how his work is explained in introductory texts—even in texts on emotion. The popular belief is that William James said, *"There's one physical expression for each emotion."*

Anyway, so I learned psychophysiology. I learned how the peripheral nervous system works and did some experiments. I did a large meta-analysis, which took a long time. Meta-analysis is just a big, statistical summary of hundreds and hundreds of studies. Well, there isn't really one physical signature in the body for each emotion.

So at that point, I thought, *Okay, it's not in the face. It's not in the body. It must be the brain.*

What led me to cognitive science, and eventually to neuroscience, was the desire to test the idea that there really was this signature set of neurons for each emotion, basically.

HB: It seems to me that there are some important philosophical aspects that should be highlighted here.

In particular there's the question of objectivity and subjectivity. Take physics. By trying to provide an explanation of some phenomenon in the world, there's a natural assumption—that few would disagree with—that the thing you're talking about objectively exists and that it's your job to provide an account of it somehow.

Meanwhile, in psychology—in fact, simply in everyday life—we've got a bunch of names for things that we observe. Some people seem to be acting this way and feeling like this and others seem to be acting that way and feeling like that.

But then to automatically conclude that, just like some elementary particle, say, those words represent something that is objectively there—there's an anger thing, there's a pride thing, and so forth—is hardly necessarily warranted.

LFB: You're describing essentialism. Is there an *essence* to anger? Is there an *essence* to sadness?

HB: Yes. And it seems to me that when you began your investigations, you had the assumption that this is the way it actually worked.

LFB: Well, that's the way that it's portrayed in the scientific literature. There has always been a debate about whether or not there are essential characteristics—either essential features or an essential element—that define each psychology category as a single thing.

But it's also true that alongside that, there have always been philosophers and scientists who have argued against that kind of essentialism. If you look at the history of philosophy of mind, the two perspectives have really just proceeded in parallel with no resolution, and usually essentialism dominates.

In psychology, we would call that "faculty" psychology: the idea that the mind is best described as a set of characteristics or abilities and each one is elemental and has some natural essence or a natural aspect to it that's basic and elemental.

I think one thing that people often don't understand—certainly about my work, maybe I haven't been clear enough, but in general, I think—is that, when you say, "*Is there something really there?*" Well, there's always something **really there**. The question is, *Is there an essence to all the phenomena that you refer to by the same name?*

In physics and in chemistry, there is a philosophical tradition of assuming that you can follow a phenomenon and reduce it down to its most basic elements and those are objectively present.

And by "objectively present", it doesn't necessarily mean that you can observe it with your own eyes—you might need fancy mathematics to prove its existence, like a quark or what have you, but nonetheless, there's something there.

There's an evolutionary biologist, Ernst Mayr, who wrote about the philosophy of biology. He wrote a lot about Darwin and Darwin's contributions to science. He made this point—which I think is very important for psychology but psychologists haven't quite embraced it yet, because nobody has actually applied it to psychology, as far as I can tell—which is that Darwin's greatest contribution is not the concept of natural selection: it's the idea that a species is a *conceptual category.*

It's not a physical category with necessary and sufficient features, and firm boundaries, and a biological essence. In fact, that's sort of the "dog-show version" of evolution: like there's a perfect cocker spaniel and all the other variations of cocker spaniels are some kind of error, because there really is a pure, Platonic form of cocker spanielness that you see in award-winning cocker spaniels who have exactly the right nose size, and exactly the right coat thickness, and so on.

One of Darwin's points was that a species is a conceptual category because the instances—of cocker spaniels, let's say—vary from one another, and that variation is *meaningful*—it's meaningfully tied to the situation, to the environment.

The idea that variability is *meaningful*, and that really all biological categories are actually concepts—because, within a category, things can vary quite a bit, and the name of the category refers to a *collection* of things, a population of things—is called "population thinking". This is why natural selection works—precisely because such variability actually exists and is important.

Meanwhile psychologists just haven't quite gotten there yet. So when I say something like, *"There's no physical signature for anger,"* people assume I'm saying, *"Anger is not real."* I'm not saying that anger is not real. I'm saying that anger is a *conceptual category*: instances of anger are highly variable, and we have to try to understand that variability and how it is that all instances of anger can all be anger while still having that variability.

HB: As you said, we know that anger exists, we know that strong emotions not only exist but they play an incredibly important role in our lives—and yet, if we're looking for a very specific, anatomical, unique, type-oriented signature, we're not going to find it. So what's going on? How do we make sense of that? My understanding is those were the sorts of questions you were asking yourself, which led to a very clear direction for your research. Is that fair?

LFB: Yes, I would say so.

For me, though, while I'm very interested in it—and my work is very influenced by it—philosophy of science came later.

I had a course in philosophy of science as a graduate student and it was one of the most difficult courses for me because I had no background in philosophy. At the time, I honestly didn't see the importance of it. In fact, philosophy of science is pretty much not taught, I would say, in most North American graduate schools in psychology, which I think is partly why we believe, for example, that we have a replication crisis.

The other day there was an article in *The New York Times* about "the failure of replication in psychology". I think there is no crisis there—the crisis is that nobody has spent much time thinking about philosophy of science. They're misunderstanding what a failure of replication is: they're assuming a Popperian framework that works in physics—although some people would even say that it doesn't work in physics anymore—but we'll talk about that later.

The use of human concepts to impose order on physical variability is not unique to psychology, but in psychology, the belief is that mental categories are respected by nature: that the brain, for example, respects distinctions like cognition, emotion, perception, and action—that they are separate parts or separate networks in the brain.

And time and time again the brain shows us that this is actually not the case.

We do have a really interesting, really cool paradox on our hands. There are philosophical tools to help us understand how it is that you can have such variability, yet anger, and sadness, and fear, can be very real things. The question is really, *What **kind** of real? What do you **mean** when you say "real"?*

Questions for Discussion:

1. To what extent is "essentialism" related to reductionism? Need this necessarily be the case?

2. Might it be argued that what Lisa is calling a "conceptual category" is simply a broader definition of what others call an "essence"? Is this a difference in kind or one of degree?

*3. In what ways does the notion of "replicating an experiment" imply the adoption of a certain philosophical framework? What does Lisa mean by a "Popperian framework" here? (Readers with a particular interest in how Popperian ideas are regularly invoked in science are referred to, among other examples: Chapter 8 of **Science and Pseudoscience** with historian of science Michael Gordin, Chapters 4 and 10 of **The Problems of Physics, Reconsidered** with Nobel Laureate Tony Leggett, and Chapter 5 of **Inflated Expectations: A Cosmological Tale** with physicist Paul Steinhardt.)*

III. Convergent Pathways

Applying conceptual knowledge

HB: There are philosophical tools, but there are also, of course, physical tools, diagnostic tools—there are experiments that one can run to build models and test those models. Tell me about your models. Tell me about the conceptual-act model of emotion and all the rest of that.

LFB: I first started off mainly as a psychologist. I wasn't thinking about the brain too much at that point—not that I wasn't thinking about it, per se, but I just knew what most people know.

Originally, the idea was that what we have are these highly variable states that vary in pleasantness and unpleasantness, or in level of arousal, and what we're doing is categorizing those using concepts that we've learned as anger, or sadness, or fear, or disgust, or what have you, based on the situation.

So if I'm feeling really unpleasant with very high arousal, that could be anger, or it could be sadness, or it could be fear, depending on what the circumstance is. This relates to a paper that I published in 2006.

There are several other theories that sound somewhat like that, but there were aspects to my theory that were unique, like the nature of concepts: a concept is not a thing; it's actually a population of instances and it's highly variable.

For example, sometimes you'll pound your fist in anger, sometimes you'll swear, sometimes you'll smile, sometimes you'll laugh, sometimes you'll tease someone.

You can do a lot of things in anger and your body will follow the action. Your body doesn't exist for having emotion; it's for keeping you alive and making sure you can perform actions.

HB: So there's no objective, Platonic anger scale that I can point to and say, "*You're a 2.8 on the anger scale.*"

LFB: No. In fact, you, Howard, when you're angry—

HB: I'm never angry.

LFB: You know, it's funny, recently my students and I realized that we're always using anger as the example and we aren't quite sure why. But everybody uses anger as the example. Perhaps it's something about Western culture.

But anyway, the point is that when you're angry it's not **one** anger—you can feel many different angers depending on the circumstance, and so can I. So can every other person who lives in a culture where anger is a concept. Maybe there is some overlap, maybe there isn't. In this case there probably is: we grew up in very similar circumstances.

When I originally formulated the theory I was coming very much from a psychological background, but then I realized that actually, if you take a neuroscience perspective, if you drop down one or two levels of analysis, you can see that the way that the brain processes information, the way that the brain even identifies what is information and what is noise, is really very consistent with this idea.

It's not just an emotion. The brain isn't just taking in sensory input and categorizing it and making meaning of it. This is actually how the brain forms *all* perceptions, *all* the time. In fact, the brain isn't "taking in" sensory information at all: that's not how the brain is wired, that's not how the brain works.

Your brain is not wired to be dormant until it's stimulated by something in the environment, which it then processes and reacts to. In fact, what's happening if you look at the anatomy of the brain, if you take a physiological approach in trying to understand metabolic efficiency, you see that what the brain is doing is *predicting* what sensory information will be coming and *comparing* those predictions to the information that actually arrives: it's forming perceptions of emotional experience, of somebody else's emotion—of everything, really—in this predictive way; and it's using prior experience, which it organizes as concepts.

If you don't have a concept for a set of sensory inputs, you'll be experientially blind to those inputs. There are these clever visual illusions that demonstrate this sort of thing very clearly. There are these black and white blobby images, and people stare at them and all they see are black and white blobby images. They're experientially blind to what's in the image. Then you can either tell them the category, or show them a photograph from which the blobby image was made, and all of a sudden, they know what to look for and they see it.

The interesting thing though, is that the blobby image is still exactly the same, but they now *see* something in that image. And for most people, now, for the rest of their lives, when they see that blobby picture, they'll see that image.

HB: Because it's part of their experience, their predictive pattern.

LFB: Yes. And they can't unsee it, or at least it's very hard. They could probably train themselves to unsee it, just as a painter can train herself to take an object and decompose it into pieces of light and then paint the pieces of light. That's what really skilled artists do, but it's an act of will. That's how virtuosic artists can make three-dimensional objects on a two-dimensional canvas.

If you took an apple and you tried to paint it as an apple on a canvas, it would look like a really crappy-looking apple. But if you take the apple and you break it into pieces of light so that what you're painting are the pieces of light—it's almost like paint-by-numbers, except you, as the perceiver, are decomposing the object into visual elements—then you can render a pretty decent three-dimensional looking apple on a two-dimensional canvas.

These illusions help us to see something about how the brain is processing information that we wouldn't normally be able to see: your brain is predicting all the time and its predictions are not abstract. They are quite—what cognitive scientists would call—"modal": the motor and sensory neurons change their firing as a consequence of these predictions.

HB: Well, that's not surprising to me. It would be very surprising if you were to say there was no neurophysiological effect and that, somehow, there was a psychological effect.

LFB: Sure. But I think the following is very surprising to most people.

Suppose I were to say to you, "*As you're looking at me. I want you to keep your eyes on my face, but I want you to imagine in your mind's eye a McIntosh Apple*". Can you do it? Can you see a McIntosh Apple? Sort of, right? You can conjure a weak image of a McIntosh Apple.

Well, what your brain just did is construct a representation of an apple, and the reason why you can see a fuzzy, filmy image is that there are actually neurons in your visual cortex which have changed their firing. But that's part of all perception. That actually is the prediction that you would have if I then actually took out an apple and showed it to you.

HB: So the same neurons that would respond if you took out an actual apple are, somehow, at least weakly, replicating that when I imagine the apple.

LFB: Yes, exactly. If I took out a real apple, a real McIntosh apple, if the apple actually looks exactly like your prediction, there will be no change in the firing of your neurons because they're already firing in that way, so your perception is completely determined by your prior knowledge stored in your head.

It's only in the case where the apple is slightly different—maybe it's a little green, maybe it's not quite shaped in the same way, maybe there's a bruise on it, or what have you—your brain is only encoding the difference between your prediction, which is represented as firing neurons in visual cortex, and this apple.

HB: This gets back to what you were asking before, as I understand it, which is, *How are these mental states constructed in terms of the brain?* In one of your papers, you talk about three different ways or three different pathways.

There's the objective perception that's coming in, there is what's happening internally in your body, but there's also all of your prior experience—your memories, your thoughts, your orientation—and all of these things are combining, not only in terms of your mental states, but neurophysiologically as well.

LFB: That's exactly right. What I would say is that your brain is not just anticipating sensory inputs from the external world. From your brain's perspective, your body is also *part* of the environment. Now, it may be a really special, important part of the environment, but from the brain's perspective, the body is just another part of the environment that it is anticipating sensory signals from, the sensory signals that you mostly don't consciously access.

We don't generally think about the body in that way. We don't have really precise sensory feelings in our body—only if something particular happens, and even then often it's usually not very exact. Take something like appendicitis: you have a vague sense of discomfort in your torso until it's almost ready to burst, and then you have a very precise pain. But for the most part, even when you run up the stairs and you feel your heart thumping, you're not so much feeling the contractions of your heart; what you're feeling is how that changes the somatosensory feelings against you're skeleton. What you're feeling is your heart thumping against your chest. You're not actually feeling the contractions of your heart.

So for the most part, we don't consciously experience what's going on in the internal environment of our body's, but from the brain's perspective, it has to keep pretty close tabs on that. So there's sensory input that comes from the body that the brain is anticipating, just like the rest of the world.

What the brain is making sense of at any given moment is the entire sensory array of this particular internal set of sensations in the context of this particular set of external sensations. *That's* what the brain has to make sense of. *How* does it make sense of it? Using past experience—knowledge, if you will—that's organized as *concepts*; and the concepts are highly variable in a given category.

That's why you can have perceptual differences in people with different, conceptual systems.

If you're an artist, say, there are many different kinds of blue that are different categories for you, that you can place perceptual boundaries around—whereas for somebody who is not an artist, all blues are the same.

Meanwhile, in some other cultures, "blue" isn't even a category. In Russian, there are two basic blues, but in English, there's only one. It's not that you can't see the variability. The question is whether you treat the variability as meaningful or whether you treat it as noise. Russians treat the variability in the two blues as meaningful, whereas we treat them all as blue.

HB: And this is strictly analogous, it seems to me, to what you were saying before about the optical illusion—whether you have the ability to see something, whether you're being trained to see something, whether, once you see it, you can't unsee it very easily without an act of will.

LFB: Right. The other thing I find really useful about the illusion is that you don't have a sense of agency: you don't feel yourself *applying* your conceptual knowledge, you can't introspect about it, and it's very hard to undo because it happens so automatically and effortlessly.

This is actually my theory of emotion: what's happening is that, every waking moment of your life, your brain is anticipating and making sense of sensory inputs from its environment—the combination of the internal environment of the body and the external environment—and it's using conceptual knowledge to do that. That's basically it.

The theory is not domain specific. What I mean by that is it's not a special theory of emotion. I'm not saying this *only* works for emotion. I'm saying this is really how the brain works: it works this way when you're constructing a visual perception, when you're constructing an auditory perception, when you're constructing an emotion, when you're thinking, and so on.

For example, the only difference between imagination and actual visual perception is that there's no prediction error. When you're doing something imaginary, there's no external input to compare your prediction to. It's just all what cognitive scientists would call simulation.

In some ways it's a very parsimonious theory, because it doesn't invent any new mechanisms or processes. It's just saying that this is how it works, and this is how it works for virtually all experience.

Questions for Discussion:

1. To what extent does Lisa's theory give us added insights into the nature of the creative process?

2. Might there be negative aspects to not being able to "unsee" something? Could Lisa's framework be fruitfully applied to conditions like PTSD?

IV. Networks

A key conceptual category

HB: Let's dig down a little bit into the theory. I think I understand the basic principles as you've described them, but I'd like to get a better idea of how things actually work in the brain, in terms of underlying networks and how these things are interacting and interrelating. Can you give me more details about that?

LFB: Sure. For a very long time, people assumed that a neuron was basically off until it was stimulated, in which case, it would send an action potential down it's axon. The reason for thinking that this is how the brain works—that the brain is basically off until it's stimulated—is that in the original studies of neurons, like in giant squid, they would take an individual squid neuron—which is big enough that one is able to study it—and they'd stimulate it, watch the action potential, and measure it.

The thing is, though, I think this is an example of the difference between physics and cognitive science—or, at least people's cartoon version of physics; I don't think this is actually how physics works—when you study an individual neuron by itself, it functions very differently than when you study a set of neurons networked together.

In what follows I'll slightly overstate things to make a point.

A network of neurons has the property of "intrinsic firing"—meaning that if you take some network and you stimulate one neuron, as long as you give that network enough oxygen and nutrients to continue, it will continue to fire forever, basically. So one neuron will stimulate the other neurons it's connected to, which will stimulate other neurons in turn, and you have what's called "intrinsic activity".

And as long as the neurons have enough metabolic stuff to keep them going, that intrinsic activity will continue forever.

What is your brain? We've gone from thinking about the brain as a big gray blob, to thinking about it as regions, to thinking of it as 100 billion neurons that are networked together.

And in this networked picture the brain has intrinsic activity: the brain doesn't need to be stimulated.

If you put somebody in a sensory deprivation chamber, they're still having sensations, they're still having thoughts, they're still having feelings, they can still have visual hallucinations and so forth. What are those visualizations? Those are predictions. Those are simulations that are not being corrected by the world because there is no visual input from the world.

So the brain has intrinsic activity, and that activity is not random—it's structured by the way that neurons are wired together. People will sometimes refer to axons as "white matter" because there's a myelin sheath that wraps around many neurons—not all of them, but many of them—to make the conductivity faster, the axon firing faster, so people refer to axons as white matter.

We would say that the intrinsic activity is structured by the white matter of the brain, by the connections of the actual neurons. This has allowed neuroscientists to discover that there is a set of what are called intrinsic networks that exist in the brain.

Now, of course, it's tempting to think of a network as a thing. If you're an essentialist, you might treat it like a physicist would treat an atom. But that's not really how it works, because even at the level of a network, the network is still made up of gazillions of individual neurons, and some of those neurons are firing some times but not at other times.

There's something called network homeostasis, which is how different neurons influence each other to cohere into a network.

For any network, the actual neurons that constitute that network might change from moment to moment, but it's the same network throughout, broadly defined.

At one level of analysis, it's a network. At another level of analysis, if you were actually looking at neurons, it might be different neurons.

So each instantiation of the network can be slightly different, yet it can still be that same network. The network is a concept and it has a population of instances all the way down.

Again, you use population thinking—Darwin's population thinking—all the way down to the level of a neuron. In fact, you can actually use it all the way down to the level of molecules and atoms.

HB: But presumably you can broadly categorize these networks: where they are, how many of them there are, and so forth, right?

LFB: Yes. But people debate about how many there are, because it depends on how finely-grained your analyses are. But what I will say about these networks is that, while of course there's individual variability and there's variability from one moment to the next, but broadly speaking there are a set of neurons that pretty much every normal functioning brain seems to have everywhere. Every time a study has been conducted to look at them, we've been able to see them.

When babies are born, they have some of those networks and not others, because other networks form with experience. The first of these networks that was discovered was named the "default mode network", because the idea was—well, it's actually a really interesting story.

The evidence for this network has been around really since neuroimaging began. In neuroimaging you're always comparing an active state—where you're presenting a subject with a stimulus or you're asking them to do some task—to some baseline. The baseline is often just looking at a fixation cross, where the brain is "at rest"—meaning you, as the experimenter, are not explicitly probing the brain of your subject with some task.

HB: Which doesn't mean, of course, that it's not being "probed" internally, say.

LFB: Exactly. If you take the view that the brain is off and it can only be activated when you stimulate it with something, it becomes really perplexing that some neural activity appears to decrease relative to the baseline. How could that be the case? For a long time, people didn't even recognize what this was. Eventually people figured out that this network is really a robust network, and they labeled it the default mode network.

Other people give it other names. Neuroscientists are just like psychologists in the sense that, if you're studying memory and you see this network, you'll call it a "memory network"; and if you're studying perception and you see this network, you'll call it a "perception network"; and if you're studying how one person thinks about another person's personality and you see this network, you'll call it the "theory of mind network", or a "mentalizing network", or an "empathy network", and so on. There are so many names for this network because everybody is naming the network in line with their own scientific interests.

The fact is, there's one network—we can call it the default mode network—and it's very easy to see. It was the first network that was discovered like this, that exists in the brain "at rest" in its default state when it's not being probed by an external stimulus—like an image—or not being asked to perform a task.

Then, very quickly, neuroscientists realized that there isn't just one of these networks; there are actually many of these networks. There have been many studies done to show that these networks are pretty stable, and at the network level they're stable—which doesn't mean that it's the same neurons every time as I was saying earlier.

So there's the default mode network. There's a network called the salience network, which is supposed to increase its activity when something is particularly relevant to you, or salient, or important. There is also, depending on how you view it, either one control network or several control networks, which influence neural firing of other networks.

HB: These are meta-network states or something like that?

LFB: Well, it's just that sometimes they break apart, sometimes they work together as one network, so what scientists are now trying to figure out is where the boundaries of these networks lie.

HB: This brings up a couple of questions. Maybe I'm going to be betraying my knee-jerk essentialist, reductionist tendencies or whatever, but it's hard to know what to do, because I want to say, "*Okay, this is where something is defined*" or, "**This** *is what we're talking about—we're talking about networks, so let's talk about a network.*"

I understand that it's variable. I understand that it doesn't always involve the same neurons. I understand that the mental states that are associated with it may arise out of all sorts of complex combinations of one network with another. I get all of that.

But I still want to know what we're talking about exactly. Why is it *this* and not *that*? Why is it *this* part of the brain, roughly speaking, and not that *part* of the brain?

LFB: We recently published a paper, building on other people's work—that's almost always the case—here involving the neuroanatomy of macaques engaged in perception.

We basically showed that there's a certain kind of brain tissue that seems to be sending predictions but not receiving them. You can make an anatomical argument—I could go into the details, but that's probably not necessary—based on existing experiments and also anatomy, that there's some tissue in the brain which sends predictions but doesn't receive predictions.

So if you think about the brain as a hierarchy, it's really at the top of the hierarchy: it's sending predictions which are then "unpacked" into their sensory details when they finally reach the primary visual region, the primary auditory region, the primary sensory and motor cortices, and so on.

So what is this tissue? Well, you can name it anatomically, or you can give it its functional name, which is limbic tissue. Limbic tissue has a particular anatomic structure, which puts it at the top of the predictive hierarchy in the brain.

It's ironic, because limbic tissue used to be thought of as the home of emotion—in fact, a lot of what limbic tissue does involves regulating your internal physical state.

This is not the only thing it does, but a large portion of limbic circuitry is partially devoted to regulating your internal environment, and at the same time it's sending the predicted sensory consequences of that change out to the rest of the brain.

People used to think of it as emotional and they used to think of it as the epitome of "reactive", but it seems that it's actually entirely predictive.

Now, where is that tissue? Well, that tissue can pretty much be found in every intrinsic network but it's very concentrated in two: the default mode network and this so-called salience network that I mentioned earlier. Those two networks contain most of the limbic tissue in the entire brain.

The default mode network is important for many things, including conceptual processing. Some of its regions are very important for memory, for example.

This is why it's a ubiquitous network and why the term "default mode" is not so bad, because without that network, your brain would not function normally.

In Alzheimer's disease, it starts to atrophy, which is why you see these really broad, pervasive deficits in Alzheimer's disease. There's even some limbic tissue in the executive control network—the network that is involved in planning, and deciding between options, and for inhibiting a prepotent response.

Scientists used to think that these networks were completely independent of each other, because they're being essentialists—which I don't mean as insulting—that's our typical way of viewing the world. Some people, like the developmental psychologist Paul Bloom, believe that babies are born as essentialists.

But the point is that people assumed that these networks were completely independent of each other, like they were little modules. But it turns out they're not, actually: they overlap with each other and they overlap in certain regions of the brain—hubs—and they actually influence each other because they share neurons.

Furthermore, the majority of these hubs—neuroscientists actually call them "rich club" hubs, because they have a lot of connections—where are they, exactly? Well, over half of them fall in the default mode network, the salience network, and the executive control network. And many of them are limbic.

So these nodes make up the backbone of communication in the brain—they issue predictions to the rest of the networks, basically, and that's how mental function occurs, including the construction of emotional experiences and emotional perceptions.

Questions for Discussion:

1. Does the notion of networks imply that there needs to be some sort of "master control" network that can ably direct all the others?

2. If the same neuron is involved in several different networks, how can it "know" when it should be firing according to one network and not another?

3. How can we be certain that there aren't other types of tissue, in addition to limbic tissue, that "send but don't receive" predictions? To what extent can the notion of a "prediction" be quantified and distinguished from other signals?

V. Slow Progress

Some slower than others

HB: I have two questions, a neuroscience question and a sociological question. My neuroscience question is: Given what you've just told me, it seems to me that, if we want to understand how his stuff works, a reasonable way to proceed is to look at some sort of large-scale network theory.

In other words, if I have n different identified networks that overlap in various ways, I'd like to be able to probe how, exactly, they are joining up to produce different epiphenomena that I can see around me, which would hopefully enable me to get some reasonably clear sense of what the mechanisms actually are in terms of how these guys interact. First of all, do you think that's a reasonable approach—and, if so, how far along are we with that?

LFB: Yes, I think it's absolutely reasonable. In fact, I would say it's imperative. What that means, is that you have to give up reductionism. It's really not the appropriate approach.

HB: Reductionism on the neuronal level. I can still look at the networks as the atoms, as it were. I mean, I can still have network theory.

LFB: Yes, but you can't look at one network in the absence of the other networks; you always have to take a holistic approach, which means you look at the functioning of an element in the context of all the other elements of the system because the resulting response or behaviour is different than if you just looked at that one network by itself.

HB: Right, because you get these emergent properties.

LFB: Exactly, you get emergent properties.

There's one other aspect to this that I haven't really talked about. Every biological system has something called degeneracy, which is a really horrible name. It's not redundancy. What it means is that there are many ways to get to the same, emergent outcome. This is not something that psychologists are acquainted with and I'm not even sure how popular it is in biology, quite frankly, but it is a biological concept.

HB: It's a pretty big concept in physics as well.

LFB: Right. But what this means is that it's not just *one* combination of networks that can produce a single instance of anger, say—there are many different combinations that could produce it.

HB: Sure, but as long as it's finite, you should be able to quantify that to build your theory, right?

LFB: Absolutely. So the first point I want to make is that you can't look at elements—regardless of what you want to call an element, whether you want to call an element a neuron, or whether you want to call it a network, or whether you want to call it the brain—without looking at it in the context of the other elements.

For example, at the level of the brain, what are the other elements? There's your body, there's what's going on in the world, and so forth.

What this means is that, if you stick somebody in an experiment where you're trying to control everything and then you just probe the subject with a stimulus, first of all, you're breaking prediction, because you're not allowing the brain to do what it normally does, and you've also removed a lot of the context.

The brain will choose other things as context. It's always using whatever it can find as context. But the results that you get will be very different than what you get in the real world where emergent phenomena are happening all the time. That's the first point that I want to make.

So when you ask, *"How far along are you?"* I would say, *"Not as far as we will be once people realize this"*, because it requires a really different way of measuring, and analyzing, and experimental design, than what we're used to, both in neuroscience and in psychology, I would say.

HB: Well, this brings me to my sociological question. Maybe you're just a particularly convincing person or maybe I'm easily convinced, but all of this seems completely reasonable to me.

In fact, I would argue that you need something along the lines of this sort of understanding to explain why we're so far away from a coherent understanding of so many things. It's obviously very complicated.

LFB: Yes, it's complicated.

HB: So you seem very convincing to me, yet I'm getting the sense that much of what you're telling me now is by no means universally recognized or accepted in psychological circles—this idea of these different networks and how we have to move beyond essentialism to far better appreciate the importance of variability within larger conceptual categories.

Secondly, just to make my question more elaborate and harder to answer, you just made a distinction between "the psychological community" and "the neuroscientific community"—maybe we can pick up on that later—but the specific question that I have now is that you said, even in the neuroscientific community, this is not, in any way, universally adopted. That's somewhat surprising to me.

LFB: I suppose it can be surprising as an onlooker, but I don't think it's surprising if you consider the history of where people are coming from. Thomas Kuhn made this point that you can't really understand a scientific field unless you understand its history. I think that's a very important point.

I will also say that history of science is also not taught in graduate programs, and it probably should be, because it's important to realize—I won't say "how revolutions happen", because I'm not sure "revolutions" ever actually happen—but how it comes to be that there's a sort of re-anchoring and readjusting away from people's basic assumptions.

If you are raised in a Western culture where you have particular embedded assumptions about how the brain works, how the mind works, and then you train as a scientist in this culture, your research will naturally be structured in a way which reflects those assumptions.

In psychology, for example, the experiment is used as a metaphor for how the mind works, because you're stimulated by sensations, those cause some kind of processing inside your brain, or mind, which is a cognition or an emotion, and then that produces some kind of a response. That's the way experiments are structured. That's the metaphor for how the mind works.

But the brain doesn't work like that.

So, why are we not as far along in psychology? I think it's because we're running experiments that were designed based on 19th-century physiology.

If you look back at the history of psychology, the field emerged as a science in the mid-19th century when philosophers, physiologists, and neurologists attempted to take the mental categories from mental philosophy and find their physical basis using the methods of physiology and neurology.

The psychology experiment is really a rendering, in a way, of 19th-century physiology: something is dormant, you stimulate it, it responds, you can then stick something complicated in the middle, like cognition, and there you have it. That's the view of the mind that's been around for a very long time and that's how experiments are designed.

Given that the model is so wrong, it's amazing how much psychologists actually *have* been able to discover. That being said, we have to start modelling variability as something other than error.

When I say "variability," I mean temporal variance: where your mind is right now is partially determined by where it was a moment ago, the state of the brain right now is partially determined by what its state was a moment ago—"state" being defined as the configuration of neural firing.

So variability temporally, but also variability by context. If you put me in one context versus some other context, I might respond differently than you would.

So there's variability at multiple levels, which are usually aggregated out as error. We're missing a lot of the information that we need to have a more comprehensive science, I would say.

HB: Right. But let me try to be more specific, because I don't think I asked a question that was focused enough.

Let's just talk about networks: the idea of networks, the idea that such things actually exist—however they're constructed and however they interact, and wherever, specifically, they may be located within the realm of variability—just that such things exist, and that they play a primary role in our brain's conjuring up the mind, if you will: is that a claim which is disputed or contested amongst psychologists today?

LFB: Most psychologists don't even think about the brain.

HB: Really?!

LFB: Well, I may be overstating the case.

HB: What do I know? You're in the field. I'm just reacting in a surprised way.

LFB: Perhaps I overstated it. But I will say this—again, just permit me to take a little bit of a historical view, because I think it's important.

In the United States, we always talk about William James, but there were other real luminaries at the time, one of whom was Wilhelm Wundt, who lived in Germany. In fact, Wundt was older than James, and many of James' ideas came from Wundt.

There's actually an interesting story about Wundt and why he wasn't as important in American psychology as James was.

HB: Well, he was German.

LFB: Exactly. The First World War made a really big difference.

HB: I just meant that he wasn't American.

LFB: But there was still a lot of flow between Germany and America back then—James went to visit Wundt and so forth. The First World War was actually a significant factor as to why he wasn't as big in America.

There were other factors too, like he was misinterpreted and mistranslated by one of his students who translated his work, Edward Titchener, who translated Wundt's work in line with his own beliefs. This goes to show that you should treat your students really well, because your legacy rests with them.

Anyway, there are lots of reasons, but my point is that Wundt advocated for a psychology without reference to biology, a psychology that had its own mechanisms and processes and could be self-contained in its explanatory structure at the level of psychology; whereas James was saying, "*No, we always have to reference back to physical things.*"

Surprisingly, although much of psychology was influenced by James, in this particular way psychology was very influenced by Wundt: many psychologists still believe that it's possible to have a science of psychology without referencing the brain or the body, frankly.

Some would acknowledge that it's important—that they have to know about it—but they nonetheless don't have to be constrained by it, that their theories can exist independently of biology.

The thinking is that if mental phenomena emerge out of biological interactions then technically speaking you shouldn't really need the biology to explain a mental state—you should be able to do the relevant analysis at the level of the emergent phenomena.

I think that's actually not the case, and there are reasons why I think it's important to consider both levels down and levels up to both constrain and inform a scientific explanation—so you have to understand biology and biochemistry, on the one hand, and sociology, anthropology, and history on the other. But that's another point.

The main point I want to make here is I that do think psychology still struggles with itself over how much attention it has to pay to biology, and to neuroscience more generally. This is an ongoing debate.

HB: Presumably it's changing. I mean, in the last 10–15 years, there are many more psychologists who are doing fMRI, among other techniques, looking at things from a physiological perspective, designing their experiments with biological and physiological factors in mind, and so forth.

LFB: In terms of a general characterization of the field, that's clearly true. But that debate, I think, is still alive and well.

But I will say this: you asked me about the question of networks, and I would say that in cognitive neuroscience, it's pretty well accepted that there are important networks, and they've slowly moved from looking at individual networks, trying to localize phenomena in individual networks in this reductionist way, to looking at the way that networks are interacting.

But in affective neuroscience, in social neuroscience—I have my foot in all of those sub-disciplines—I think they're a little slower. It's not that they don't believe it per se, it's more that they don't know it. It takes some years for knowledge to percolate from one discipline to another. It's not ignorance in the colloquial sense; it's ignorance in the definitional sense: many people simply aren't aware.

HB: And there must be practical issues as well, I would imagine. Just off the top of my head, to take some examples: when I understand that you're doing work on the interchange or interaction between emotions and vision or emotions and linguistics or language, this makes sense, it seems to me, from what you were saying before about these networks—because the networks don't respect these categories; the networks are just dealing with all sorts of things that are going on, so there's a natural overlap. You're looking at it from a more holistic perspective and you're probing exactly how these networks work and how they interact using these different tools.

But if you're coming at it as somebody who's just been looking at, say, facial nerve signalling, or you're just looking at it from the perspective of physiological signs of anger, because that's what your training is and that's where you come from, it's naturally difficult, I would imagine, to start coordinating that. Is that reasonable?

LFB: I think so. I guess there are two ways to answer that. One is to say that, even in my own work, I'm not really testing the hypotheses in the way that I'd like to, because I don't have the methods that I need.

HB: What do you need?

LFB: Well, for example, I would need an ambulatory setup that would measure a subject's brain activity and their physiological activity continuous with videotaping their environment and ambulatory eye-tracking, which would allow me to know what they're looking at; and all of that occurring at the same time as they're occasionally probed for their subjective experience—and all of that is streamed in real time, so that you can get anywhere, from one signal to any other.

That doesn't exist. We're trying to build that, but that doesn't exist. I've been working with some engineers at Northeastern for several years to come up with an analytic framework to analyze data that way, to analyze all channels simultaneously, which is really what I think we need in order to integrate things properly. It sounds sort of like a grab bag, like, "*Measure everything*," but it's not really.

If you take this kind of approach, where you're not a priori determining what is signal and what is noise; where, rather, you're letting the data tell you what is a signal and what is noise—

HB: Then you have to record everything, presumably. You have to.

LFB: Yes. So I think one answer to your question is that the methods aren't really available yet, even in neuroscience.

It's remarkable what we *can* do, but what we would really need to do in order to properly answer these questions is far beyond what people are typically doing. That's not a criticism. Things are happening really fast, and maybe in 5 or 10 years we'll be there. I think that's one way to answer your question.

But I think there's something more that's necessary. It's not just experiments and analytic methods and so on—we really need a shift in how we *think* about phenomena. We need a theoretical shift, and I think we need a philosophy of science shift. We can't continue to treat phenomena as elemental, when they're not.

In psychology, there's a tendency, in my experience, to confuse a phenomenon with a process, or with a mechanism.

For over a hundred years now, there's been this tendency to make this false correspondence between phenomenon and process: if somebody experiences a threat, say, then there must be a "threat detector" in the brain, and if somebody is experiencing anger, there must be an "anger circuit" in the brain. There's this basic conviction of a 1:1 map between the mental phenomenon and some kind of more basic mechanism. But that's just not how it works.

I think neuroscientists are a little further ahead as a general group, but some parts of neuroscience are much further ahead because they have the advantage of seeing things that other people can't see.

Take the anatomy of the brain. If you look at the structure of the brain, it's clear that it's not set up to be stimulus-driven, but unless you actually read anatomy papers you just wouldn't know that. And you also wouldn't know, for example, that, metabolically, a human body can't support a brain that *is* stimulus-driven—it's just too metabolically expensive. But again, you wouldn't know that unless you read these metabolic function papers about the brain and its relationship to the rest of the metabolic economy of the human body. You just wouldn't know.

HB: When you say "stimulus-driven," do you mean *exclusively* stimulus-driven? Do you mean in terms of energy balance? Because I *do* have these visual networks that are going on in my brain, which *are* stimulus-driven, presumably.

LFB: Well, yes and no. Let me clarify what I mean. Is it the case that sometimes you don't predict something and you are, therefore, captured by a stimulus? Sure, absolutely. The brain sometimes doesn't predict well and sometimes we seek and cultivate circumstances where it doesn't predict well—that's what novelty is, that's what learning is.

Learning is where you're not predicting particularly well and you have to acquire some new knowledge in order to be able to do it.

Babies are born with some conceptual machinery—how much is a matter of huge debate—but for my purposes, let's just say that they're born with some goodies, but we don't know which ones. But for the most part, they're really engaged in "prediction-error" almost all the time—not always, but largely.

This corresponds to what Alison Gopnik, the developmental psychologist, calls the "lantern of attention": it's not like a spotlight like it is with an adult, because much of what babies are learning is novel to them, so they're not predicting particularly well; and interestingly enough, there's no default mode network in infants. They're not born with one. It develops over the first couple of years of life—exactly when the basis, the foundations of conceptual development are occurring.

HB: What years does it develop?

LFB: Well, that's also a matter of debate. Some people say by the age of two or four, some say by the age of eight or ten, but something that starts to look like a default mode network starts to develop somewhere in early childhood. I think we'll have a better answer to that really soon. This is all very new stuff. It's only in the last 10 years or so that people have really started to focus in on these intrinsic networks, so we still have a lot to learn.

My point is that knowing certain things about the networks—like how they work, or knowing about metabolism and how it functions—constrains certain kinds of explanations at the psychological level. Some become plausible and some become largely implausible. One thing that's implausible is this idea that your brain is dormant and is then stimulated.

Some people think that we're using Bayesian-type probability, but whatever the details are, your brain is estimating, based on the sensory array right now, what the sensory array will be in a moment from now. So, is it stimulus-driven? Sure, but not in the way that people typically mean it. Is it the case that we sometimes can't anticipate the sensory input so we're surprised, or something is novel, or we're horrified, or whatever? Sure, absolutely. And do we sometimes seek those things? Absolutely.

When you go to a new culture and you've never been there before, everything is miraculous because you can't anticipate it. When you learn a new language, it all sounds like gobbledygook, until you have enough experience that you can start to place boundaries between words and so on.

HB: Right. But that doesn't mean that you're not predicting in the meantime—you're just predicting and failing.

LFB: Exactly, and you can think of a failure to predict in different ways. There are now a couple of papers that look at something like autism, for example, as a failure of prediction.

The idea is that abstract concepts are products of how the brain is wired—the brain has a gradient, let's say.

The primary sensory regions at the back of the brain are filled with a lot of neurons and relatively few connections between those neurons; it's not that there are no connections—they have cortical columns where the neurons are connected—but there are fewer connections and more little neurons.

Meanwhile at the front of the brain, there are fewer neurons, but the ones that are there are much bigger and there more connections.

So there's this gradient in the brain that's effectively compressing information. Prediction error comes in and it's like it's being compressed from its sensory particulars into summaries.

But what if your brain wasn't able to do that because you had some kind of physical dysfunction, either with limbic circuitry or with connections at the front of the brain, and you were left with these highly detailed representations, and you couldn't generalize from one instance to the next? You would have real problems predicting— you could predict only on the basis of exact matches to the current situation.

That's not how a normal functioning brain is predicting; it has all this information which it can combine in interesting Bayesian ways and so forth. But what if you couldn't do that and you had to basically pattern match specifically, situation to situation, in order to predict?

There are these really fascinating autobiographies of individuals who live with autism where this is exactly what they describe. The ones who are articulate enough to introspect about what is happening explicitly say, "*What I try to do is remember a situation which is **exactly like** the one that I'm in and then I know what it means and then I know what to do.*" But that's not how a normal functioning brain works.

Questions for Discussions:

1. *What does Lisa mean, exactly, when she says that we have to "give up reductionism because it's not the right approach"? What is reductionism, exactly? Additional Ideas Roadshow references on the limitations of reductionism occur in Chapter 4 of* **Minds and Machines** *with neuroscientist Miguel Nicolelis, Chapter 3 of* **Pushing the Boundaries** *with renowned scientific polymath Freeman Dyson and Chapter 7 of* **The Problems of Physics, Reconsidered** *with Physics Nobel Laureate Tony Leggett.*

2. *Do you think there will come a time when all psychologists will appreciate the importance of adopting a rigorous biological understanding in order to comprehend the mind?*

VI. Towards Genuine Dialogue
The benefits of interacting with essentialists

HB: I'm guessing we're a long way away from this, but can you imagine any specific physiological signatures corresponding to these ideas, something like saying, "If this and this happened then we would have a failure in this network—or series of networks, shape of a gradient or whatever—that would manifest itself in that particular way". Might there be something specific that could be envisioned along these lines?

LFB: I think so—there definitely is. But I also think that we have to be aware of essentialism, because it really is problematic, certainly in psychology and neuroscience, but I think just more generally as well.

In the past, like the 19th century, a "holistic approach for the brain" meant that every neuron could do everything—which is called equal potentiality. But that's not what I mean when I say holistic.

Here's an example of what I mean. One of the things that Darwin conceived of was population thinking: this idea that a species is a conceptual category, variability is real, and so any summary of the category is an abstract representation or summary of the category—it doesn't even have to exist in nature.

It's like saying, *"If you have a bunch of instances and you take the mean, the mean doesn't even have to exist in the population,"*—the mean is just a mathematical relationship that describes, roughly, what people are experiencing. For example, not every middle-class couple has 3.3 children—in fact, none of them do. So the same logic would apply.

For example, we took somewhere between 80 and 300 different studies of emotion—I can't remember exactly how many it was right now—and we did something called pattern classification, where you basically train a mathematical equation to try to distinguish one emotion category from another.

So you know what the real category is for the study—let's say anger is being cultivated in one particular study, and in another study it's sadness, and in another study it's happiness, and so on—you train a classifier to find the pattern that is the summary for each category, and then you can use that to predict new instances.

You can do it. We did it, and we published it. In fact, several people have done individual studies where they have used individual subjects and found classifiers that can distinguish anger, sadness, and fear, in a given subject, and predict it in new subjects.

The problem is that then people understand the pattern, not as an abstract summary, which doesn't have to exist in *any* of the instances—just like those 3.3. children—but they understand it as the *essence* of the category.

In other words, they believe that this pattern will exist in every instance of the category and they think that this *is* the biological essence of anger, or sadness, or fear. And if you look at the way that most of those papers are written, that's what they say. And if you look at how the media report these studies, that's what *they* say.

But, in fact, this is evidence of populations: each category is a population of instances and the pattern is an abstract representation which need not actually appear in *any* of the instances.

Is it possible to do science this way? Absolutely. For example, a few moments ago I described an ideal type of ambulatory setup— what if you could develop classifiers based on somebody's entire data associated with such a setup? You collect a huge amount of data associated with one subject, you apply big data techniques to it and then you use classifiers to try to distinguish one category of mental state from another.

And then, for example, you could try to see empirically how well a repertoire of classifiers for one subject works on another subject, how it maps over. So it's definitely possible to do this work without assuming that variability is error, by using population thinking, without engaging in essentialism, and using a holistic approach as opposed to a reductionist approach.

I absolutely think it's possible, and I really think that's the direction where things are heading, and I think that the conceptual-act theory, to some extent, is not the only example of a theory which tries to take account of these phenomena, but it's trying to do it at the level of the brain, at the level of the nervous system, at the level of psychology, and really at the level of cultures, which could be sociology or it could be anthropology, but it's usually what's considered social construction.

I started just at the level of psychological construction, but it's really a full-range constructionist approach, from biology all the way up to something that would be closer to social construction. The difference, I think, is that this theory, unlike other constructionist approaches, is really grounded in biology.

In the past, an essentialist approach, which has always been linked with evolution and biology, has been in contrast to social construction approaches, which are linked with none of those things. And some social constructionists have said explicitly that those things are not important, and others just haven't incorporated those things into their theories, so it's not hard to believe that they think it's not important either.

But, as far as I can see, there have not been very many theories that are attempting to use a constructionist approach with population thinking and all the other aspects I've mentioned in a gradient of levels of analysis. I think that's the distinctive piece: that I'm infusing construction with concepts from evolution and biology, in particular, which give it actual plausibility as a scientific framework, that I don't think it had before.

HB: What would people of a very different mindset say? What would their response to this be? That it's too complicated—intractable, maybe? Or something else? Maybe I'm just naive, but I'm having a hard time imagining that people out there, be they constructionist-oriented or otherwise, would not recognize the importance of the biological world.

LFB: Absolutely. Well, I think there are some psychologists who want to say that biology is a different level of analysis, so we don't need to care about it, but I think most people would agree it's important, even if you don't do it yourself, you need to be aware of it. But I think a lot of people still think that essentialism is a reasonable way of thinking about the natural world.

HB: I guess what I'm trying to explore is, do they really believe that, insofar as they understand what essentialism is and they understand the philosophical assumptions that they're making, or do they just do it in an unthinking, knee-jerk sort of way? Do you understand my question?

LFB: I totally understand what you're saying. I don't think you can sum up a whole field with a single statement. I think that many people don't realize the assumptions they're making.

Whenever I have a new crop of students, I always tell them, *"When you're reading stuff, some of it you're going to love straight away. You'll think it's brilliant. But sometimes you're going to read stuff and think, 'This is bullshit. I don't believe this'. But, in both cases, what you've read has touched on some very basic assumption that you have that you probably didn't know that you had. So you can use your affective response as a tool to explore your own assumptions."* This is the clinician in me coming out a little bit.

I think that what happens with a lot of people is that they don't realize that they're using essentialist assumptions, but when they read my work or work like this, they just have a strong, negative reaction to it and they don't really know why.

And later, if there's an opportunity to explain my position to them in detail, it typically goes very well.

But I think that some people really *do* believe that—they're extremely intelligent and they're extremely well read, and they really believe that essentialism is a reasonable way to describe the way the world works. And, for them, this is clearly not the right way to go.

Obviously I think this is the right way to go, or I wouldn't have devoted my whole career to it. On the other hand, it is interesting to me that, for the last 3,000 years, there have been these two ways of viewing how the mind works which have really been at war with each other, and there has never been a resolution. I'd like to think that now we have better tools so there will be a resolution, but that's probably what people thought 100 years ago.

HB: It doesn't mean that it won't happen though. It might happen 300 years from now.

LFB: That's true, but I guess what I'm saying is that one thing that would be really helpful would be if the people who are really thoughtful about essentialism and genuinely defend it would work with the people who don't and are taking a more constructionist approach.

It's not that you have to have find a compromise, but I am reticent to completely dismiss a line of thinking that's been around for 3,000 years. Maybe we should understand that line of thinking as something that the brain produces. Maybe there's something about our wiring that really leads us to be essentialist. I actually think that's what's happening. I think that, because of how the brain categorizes, because of how it forms concepts, because of how it's structured, because of the gradients within the cortical sheet itself, I think that essentialism may be the way that we go about things.

But, on the other hand, there are many cultures that aren't strongly essentialist. In some cultures, cognition, emotion and perception are not the basic categories of the mind.

HB: So maybe our brains are structured with a certain potential, a certain likelihood to become essentialist, depending on what the prevailing circumstances are.

LFB: Yes, so if you look at Buddhist philosophy of mind, for example, you see a very similar debate between essentialism and constructionism, except it happened 2,000 years ago and it's still going on.

The dharmas are the essential bits of experience, and we just have to strip away—this is the sort of classical view—our illusory self and all of our attachments to an illusory self and then we can get to the true elements of experience, which are these dharmas, these little essences.

Then, several hundred years later, some other Buddhist philosopher came along and said, "*No, these dharmas, they themselves are just concepts. They're also the production of the human mind,*" which is a constructionist view. That same debate is occurring in a completely different cultural context, which leads me to think that there's something really important here for us to understand.

Maybe there are real lessons about the way the world works in essentialism, but I tend to think that essentialism stems from how the brain is processing information. Sometimes variability is not useful in a given instance, but that's a far cry from relegating it to being irrelevant as error, all the time.

But I think one thing that really needs to happen is that there has to be a real dialogue about this, not where different sides are caricaturing each other, but a dialogue where they are actually learning from each other what the key experiments need to be in order to move forward.

Questions for Discussion:

1. Is it possible that our brains are structurally unable to understand themselves? How might we know for certain?

2. Why do you think Lisa is continually striving to examine other cultural contexts to frame these core concepts? Do genuine opportunities exist for meaningful collaborations between for example cultural anthropologists, historians, and cognitive scientists? Those interested in this question are referred to Chapter 6 of **Embracing Complexity** *with renowned Princeton University historian David Cannadine, where he specifically suggests increased contact between historians and neuroscientists.*

VII. Final Thoughts

Some philosophical reflections

HB: I have a final, or perhaps a penultimate question, and it's a two-parter. I can imagine that one can be a constructionist and still believe within that paradigm that it doesn't actually make a lot of sense to distinguish between the mind and the brain once we know enough.

That is to say, I believe that it's logically conceivable that one can build such a framework and say, "*You have to take a constructionist approach, you have to take these networks, you have to appreciate variability, you have to do all of that,*" and yet, at some meta-level of understanding, there's an awareness that the mind will eventually be regarded as a superfluous category.

LFB: Absolutely not. I completely disagree.

HB: Okay, so that was my first question. I know you personally don't believe that, but I was wondering if you thought it was logically possible to adopt a constructionist philosophy and still adhere to such a position, but it seems that you believe that its negation is a logical outcome of having a constructionist philosophy.

LFB: Yes, I think a constructionist philosophy implies emergence. If you believe that the mind is an emergent property of a brain in context—a brain in a body that is moving around in a world—then the mind is something more than the sum of its parts: you can't reduce it to any part. You can't reduce the mind to the brain.

HB: OK. But if you take a God's-eye view and you say, "*Okay, I know everything you've experienced. I know the current state of individual* *neurons, and I appreciate all the networks that are actually involved. I know what you're being subjected to in terms of your sense data and all the rest of that,"* and so forth—I realize this is a philosophical discussion...

LFB: Of course. I would say, first of all, that you're setting up implausible conditions, because you can't know.

HB: That's fair enough.

LFB: Neural activity is stochastic. There's a stochastic aspect to it that we haven't talked about. So you actually can't know.

HB: See, I have a problem with that too.

LFB: Okay, but then there's also degeneracy. You say, *"If you could know everything,"* and I don't think you *can* know everything, not because of some mystical something-or-other, I think you just can't. There's too much going on to really know everything.

I'll also say that, while there are lots of individual differences in brains, once we really understand how the brain works it's probably going to work that way for most people. What will change will be the concepts that the brain has bootstrapped into itself based on its ongoing experience. That's where categories of mind come from, that's what they are: they're concept-driven meanings of the sensory array.

The description of a mind can change, the content of experience can change from one culture to another, even when the structure and function of the brain is still pretty stable across cultures.

HB: Sure, I understand that. I also greatly appreciate the practical value of this in terms of different categorizations, different levels of understanding, as you just pointed out, the broad-based similarities between individuals, the functional aspect of this, and so forth.

I guess I'm asking the same sort of philosophical question that John Searle and similarly-oriented people are asking—or at least what I interpret them to be asking—insofar as, however complicated it is and whatever level of emergence might come out, whether or not there's a philosophical objection to the notion of mapping "brain states" to the mind, as opposed to a practical one involving feasibility.

LFB: I think that embedded in your question are really two separate questions. Maybe I'm misunderstanding, but I think one question is, "*If you knew every single neural pattern, in context, and you could measure and capture them all, would you then know what mental state a person would be in at a given point in time?*" And my answer is yes. I don't think you can ever know that, but if in principle you could, then I think the answer is yes. I think stochastic influence is important, though you may not. But I think that's one question.

The other part of the question you're asking seems to be, "*Is the concept of a mind superfluous once we understand how the brain works completely?*" Maybe I misunderstood, but it seems like you're asking, "*Can you get rid of the mind if you completely understand the brain and all of its patterns and so forth?*" And I think no, you can't.

I'm not an expert on Searle, but I think there are other things to say about this. Searle's concepts of social reality—which we really haven't touched on in this conversation—are hugely important to the theory that I've crafted. That is, you and I and the cameraman and everybody else in this culture have collective intentionality about certain things that we categorize in certain ways, such that, when we do it collectively and we all agree, we actually create a reality that wasn't there before.

HB: That's right. Those things have some ontological status.

LFB: Absolutely. Money is the perfect example, and that's the example that I often use, although I sometimes use flowers and weeds, and I sometimes use cupcakes and muffins—there are lots of examples you can use. But if the mind is really emergent from the brain in functioning in context—and often, the context is other brains of other people—then you can't reduce one to the other. You'll never be able to, under any circumstances, be able to reduce. What you might be able to do is predict, but you can't reduce.

HB: You mentioned the word "stochastic" a couple times, so I should probably say a few words about this a little bit before I let you go. Let me be clear about what my problem with stochastic stuff is. Other than quantum mechanics, so far as I know, "stochastic" essentially seems to mean that we just don't understand it enough—indeed, maybe we don't even understand it in principle.

For example, in thermodynamics, we'll never be able to tag every single atom in a box or whatever it is, but that's not really relevant when I go to my God's-eye view of "*What if I could know all of these things?*" So, from a philosophical perspective, I'm just not persuaded by any stochastic arguments.

LFB: I would agree with your sociological observation. I think words like 'stochastic' or even 'emergence' are often used as black boxes, like "*something happens here that we don't really understand, and then there's an outcome.*" I completely agree with you.

That being said, just because many people use it incorrectly, or use it as a crutch or a stopgap, doesn't mean that it isn't real.

Here's one example. I'm not an expert on Shannon's information theory, but one thing I do understand a little bit is that random variation is really important to the functioning of any system that can carry information, and there are some nuclei in some animals that are completely dedicated to introducing random variability into neural patterning.

HB: OK, but now you're using aspects of quantum theory, because that's where that comes from. I said "aside from that".

LFB: Yes, but you can't say, "*Aside from that,*" because that's actually part of it.

HB: Okay, fair enough. So you gave me an unfortunate answer.

LFB: Unfortunate?

HB: Unfortunate because I don't want to get into the foundations of quantum theory, which is what we would have to do to even try to have a meaningful conversation here. And we'd more than likely fail miserably.

LFB: Right. So that's beyond my pay grade. I know some really smart physicists who could probably have that conversation with you.

HB: Well, anyone could have that conversation, but I'm pretty certain that nobody knows the answer—in fact, I'm pretty certain that nobody even knows how to frame the right sorts of questions.

But I've drifted considerably—I do tend to do that, I'm afraid. At any rate, I've really enjoyed myself, but before we wrap things up I should ask you if there's anything you feel we've missed or if there's something you'd like to add.

LFB: The only thing I probably would have wanted to talk a little more about—not that anything we talked about wasn't important, because this was really fun—is to emphasize that the power of social reality and explaining the biology of how social reality works is really an important part of construction, at least my version of construction. But I don't think that's something we can do in three minutes, but just earmarking it for people so that they can go and read about it is probably enough, I think, for now.

HB: So, since you're earmarking, would you earmark specific things? Would you like to point them in specific directions?

LFB: I always start my students with Searle, not because he's talking about biology, but because his account of social reality, I think, is really useful. I also like Ian Hacking, so I like to put the two of them together, and then I like to combine that with some of my own writing on the neuroscience of how social reality emerges, since that's really what I think mental categories are. They're forms of social reality for making sense of physical variability. There's one paper in particular that I wrote called *"Emotions Are Real"*, which I think is a useful starting place for that.

HB: Well, thank you very much, Lisa, this was enormous fun. I really enjoyed myself.

LFB: My pleasure. This was great.

Questions for Discussion:

1. What, exactly, do you think Lisa means when she talks about "creating reality" with certain shared concepts such as money? Those interested in a more detailed discussion of these ideas are referred to Chapters 3–8 of **The Social World, Reexamined** *with Tufts University philosopher Brian Epstein and Chapter 7 of* **How Social Science Creates the World** *with UC Berkeley political theorist Mark Bevir.*

2. What do you think the field of psychology will look like in 50 years?

Continuing the Conversation

Readers are encouraged to read Lisa's books, *How Emotions Are Made: The Secret Life of the Brain* and *Seven and a Half Lessons About the Brain* which go into further detail about many of the issues discussed in this conversation.

Knowing One's Place

Space and the Brain

A conversation with Jennifer Groh

Introduction

Framing Evolution

Part of the problem of appreciating the full complexity of neuro-science is that even the complexity is, well, pretty darned complex, coming in all sorts of mesmerizing varieties.

There is the obvious, "in your face", sort of complexity, based upon the sheer volume of possibilities: coming to terms with the vast number of neurons in the brain, grappling with how to go about isolating one specific protein out of thousands, and doggedly following each one down its own particular thousand-fold biochemical pathways.

But then there is a much more subtle sort of complexity, the sort that is not immediately obvious to a non-specialist but upon further reflection presents a vast number of seemingly insuperable obstacles to our understanding.

Take vision. Nothing could be more natural than blithely declaring that sight involves the brain appropriately processing the signals that it gets from our eyes. Which is true, of course, as far as it goes. But dig a little bit deeper and a much more sophisticated picture starts to emerge.

Duke neuroscientist Jennifer Groh has spent the vast majority of her research career doing just that: looking to unravel the subtle, and often overlooked, complexity of how our brains develop an under-standing of where we are.

"The photoreceptors in our eyes give us a representation of where visual information is, where objects are in the world; and that frame of reference depends on where the stimuli are with respect to the

array of photoreceptors. In other words, it depends on where the stimuli are with respect to your eyes.

*"Well, we can **move** our eyes; and we **do**. In fact, we move them a lot—about three times per second—and we move them really fast, at a speed of about 500 degrees per second.*

"That's a lot of eye motion that the brain has to deal with, to compensate for. It has to assemble the snapshots that are taken by the photoreceptor array at each of the different positions that your eyes might be looking."

That's complicated enough, but that's only the beginning. Now think about the way the brain integrates a wide range of other sensory input—vision, hearing and touch—each sense depending on completely different mechanisms, and each one having, as Jennifer describes it, its own neurological "frame of reference".

"If you then extend this problem to include some of the other senses, like the auditory system, it's important to first recognize that, of course, sounds aren't affected by how the eyes are moving. The auditory system is using a different frame of reference for figuring out where the sounds are located, which is based on subtle cues that are different across the two ears.

"A sound that's located on one particular side will arrive in that ear first and will be slightly louder in that ear than the other. The brain has to compare the signals arriving in one ear with the signals arriving in the other to compute the angle that the sound is coming from.

"In general, then, the auditory system is computing sound location based on cues that are fundamentally anchored to the head, while the visual system is computing visual locations based on cues that are fundamentally anchored to the orientation of the eyes. So every time your eyes move, you're yanking your visual scene around to some new position with respect to your auditory scene.

*"I've been really interested in how the brain **fixes** that, how it puts those two signals into a common frame of reference so that you can*

*do things like use lip-reading cues to help you understand what
someone is saying."*

Well, alright, you might say. It's complicated. Very, very complicated,
even—and perhaps unexpectedly so. But after all, when it comes to
the brain, lots of things are complicated—speech, language, regulating our emotions, playing the cello. Isn't this just yet another in a
long line of highly complex neural processing?

Perhaps. But then again, maybe how we process sensory information about the world around us is somehow different. Maybe the
neural infrastructure responsible for how the brain represents space
actually plays a more preeminent role than we might naively think.

In a knowing wink to the decidedly less rigorous side of popular
neuroscience, Jennifer begins her captivating book, *Making Space:
How the Brain Knows Where Things Are*, with a little tongue-in-cheek
declaration. *"Nine-tenths of your brain power is spent figuring out
where things are,"* she announces boldly, before immediately admit-
ting that she just made that up. Why would she do such a thing?

> *"As you probably know, there is this popular myth out there that
> we only use 10% our brains. None of us actually know where that
> number comes from, so it's kind of a running joke in neuroscience to
> say, 'I'm just going to throw a number out there and say this is how
> much of your brain is involved with this, that, or the other thing.'*

> *"But I'm half serious—or maybe a little bit more than half serious—
> because, when you look at it, there's an awful lot of the brain that has
> been identified as carrying some kind of information that's relevant
> to these kinds of processes. There's a lot of the brain that responds
> to visual information, there's a lot that responds to sound, there's a
> lot that responds to touch, there's a lot that's involved in controlling
> movements that are essential to understanding how to combine
> information across these different sensory systems.*

> *"If you were to say that all of those brain structures are really just
> doing those things, that it's their job to work on these spatial-pro-
> cessing, sensory, motor-control issues, there wouldn't be that much*

left for doing the things that concern, say, what makes us smart. How come most of the brain isn't involved in, say, language?

"It turns out that, if you look at areas of the brain that seem to be involved in, say, language, or memory, or attention, or planning, or motivation, there's a lot of overlap between the structures that are implicated in those processes and the structures that are implicated in sensory and motor processing."

Why should such overlaps exist at all? Why, for example, is there such a well-documented link between memory and space? What could be causing that?

Suffice it to say that nobody knows for sure, but Jennifer speculates that it might be intimately tied with the basic principles of how evolutionary processes work.

"A general problem in evolution is to envision how simple events, like the mutation of an individual gene, can produce an organism that functions better than the other organisms that don't have that mutation—because usually, when you tweak something, you make it worse. Intermediate states in the course of evolution are often hard to envision.

"One thing that may be happening is that modules in the brain might be duplicated through a fairly simple set of mutations so that you might take a structure that's working well, and maybe one small change means that you now have two of those. And if you now have two, and the one before was sufficient, you find yourself with an extra that can be used for something that you weren't originally doing.

"The thought is that perhaps there's still some history that the duplicated module retains based on where it came from—the circuitry might look similar to what's present in the original area because perhaps it's getting some of the same kinds of inputs, but it wouldn't be doing the same exact things. It would be doing something similar, but related to a different type of input.

"And it may be that spatial processing originally arises as something that's essential for that first module to do, and that when it gets

duplicated in that second module, you still have all this spatial infra-
structure, only now you're going to use it to do things like think and
maybe reason about abstract concepts that might easily be equatable
to something spatial, but aren't, in and of themselves, spatial."

The reason why so much of our neural architecture might seem "space-like", then, is that nature has used our spatial processing systems as a sort of evolutionary template, somehow leveraging them towards the creation of a multitude of other high-level systems that can be engaged in things like logic and language.

That sounds pretty complicated, too, I must admit.

But that doesn't mean it isn't true.

The Conversation

I. From Ticks to Brains

Becoming a neuroscientist

HB: I'd like to start by talking about your scientific motivations—not necessarily regarding neuroscience, although I'd like to get there eventually—but just in terms of your general interest in science. Did you think to yourself, *I'm going to be a scientist*, ever since you were a little girl?

JG: I can trace it back to about age 12 when I saw a program on NOVA about the brain. It had three-dimensional views of neurons and little twinkling lights travelling along the axons of the neurons and all that. I just remember thinking, *That's really cool. That's what I want to do.*

HB: So you were particularly attracted to neuroscience.

JG: That's right. But when I got to college, I thought, *Well, I was only 12 when I developed this interest. I should make sure that I try other things.* I was ready to change my mind if something else came along.

HB: And did you actively try other things? You went to Princeton for your undergraduate work, right?

JG: I did, yes.

HB: Were you in a neuroscience-oriented program right from the get-go—biology, presumably?

JG: Well, the biology program was broader than that. At Princeton everyone does a senior thesis, which you can get started on well before your senior year. I got really interested in behavioural ecology:

how animal behaviour relates to the ecological constraints that the animals are operating under.

I became involved in a project studying wild horses, examining what behavioural strategies they employed to enhance their ability to survive and reproduce.

HB: What are some of those strategies?

JG: Horses, like most animals, have an equal sex ratio at birth—an equal number of males and females. But the social structure is that there's usually one male associated with a group of females. So by the time they get to adulthood, there are naturally some males that are left out. These are called the bachelor males. I was interested in whether or not they were forming groups themselves—fraternities, if you will—of "left-out male horses"; and, if so, what advantages they were obtaining.

HB: So what did you find out?

JG: Well, I didn't stay in the field, so I can't say for sure how it turned out after that. But it looked like when they were in groups they were able to go to parts of the environment where the food was better and maybe hold their own a little bit more against the harem stallions, those males who are grouped with multiple females. The bachelor males were able to hold their own against the stronger males a little bit more successfully when they were congregating in groups.

HB: Do these bachelor males ever take on the harem stallions? Do they ever think, *Well, there are 10 of us and there's only one guy with all the mares. We can push him away?*

JG: If they did, I didn't see it. They probably do, because sometimes there's turnover; and you'll find that a male who used to be in charge of the group is no longer in this position and somebody else has come in.

In fact, I shouldn't say they're really "in charge". It's one male and multiple females, but it looks like the females are really driving the show. They're the ones deciding where to go, what to eat, and so forth, while the male is more or less just tagging along, trying to keep the other males away.

HB: Interesting. Anyway, behavioural ecology was a real interest of yours, but then at some point you decided that it wasn't as interesting as neuroscience. Tell me how that happened.

JG: Well, doing that kind of fieldwork is really hard. You're camping on a barrier island, you're living in a tent, and you're spending 10 or 12 hours a day out in the hot sun.

HB: Where did all this happen?

JG: This was on the Outer Banks of North Carolina.

HB: There are wild horses on the Outer Banks?

JG: Yes, there are about a hundred wild horses on this particular island. They've been there since Spanish galleons were shipwrecked. It's beautiful out there.

But I'd be sitting there on my little stool with my notebook, and between taking notes, I'd look down at my legs and start picking off all these ticks, which was pretty unpleasant. So I thought to myself, *Hmm, the brain is really fascinating. I think I'll go back to that.*

HB: Sounds understandable. And at the time, did you have any particular research inclination with respect to neuroscience, or was it more a case of just being generally fascinated by the brain?

JG: As it happens, the first things that I learned about were the things that really grabbed my attention, but because they were first I wasn't sure that I wasn't going to be equally interested in other topics. So I sort of circled around and tried a few other things before coming back to what I was first interested in, which was how space is represented in the brain: how we know where stimuli are located, how we reconcile information about where visual stimuli are located with information about where sounds and things touching you on your body are located.

I was fascinated by how your brain puts together all that different kind of information to give you a coherent sense of the world, a sense that you're moving around through one environment, independent of whatever the underlying sensory detection systems might be.

Questions for Discussion:

1. How influential do you think science-oriented films or television shows, such as the one Jennifer describes in this chapter, are to motivating children to become scientists? Should we have more of them?

2. Are scientists born or made?

II. Historical Background

On the shoulders of giants

HB: This brings us to your book, *Making Space: How the Brain Knows Where Things Are*, which I understand is based upon courses that you've been teaching here at Duke.

JG: That's right. I teach about perception and the neural basis of perception, and I focus on the spatial aspects of perception.

HB: One of the things that I really enjoyed about it is that it has all sorts of scientific detail in it. I should emphasize that it's a book that's deliberately written for a general audience, but it nevertheless manages to do more than just talk vaguely in a high-level way about concepts—which, to be honest, is what I had expected.

In particular, the book forced me to confront all sorts of things that I had previously not thought about, concepts that I had taken for granted. Right at the very beginning, for example, you talk about how we actually see things.

I knew that we have photoreceptors in the eyes—which is roughly equivalent, I now recognize, to saying that our eyes are the things that we use to see. In other words, light comes in and impinges on our eyes, and those photons get converted to a current, and that goes to the brain, and that's how we see. That was what I knew before I opened your book, and I thought that was good enough to say that, more or less, I understood how "vision works".

But suddenly, upon reading your book, it occurred to me that I hadn't paid any attention whatsoever to any of the vital underlying processes that go on at the molecular level to explain all of this.

You talk about how a photon comes in and creates the rotation of some molecule by breaking some carbon bonds so that another chemical process can now be unblocked. None of that, I realized, had ever even registered in terms of an awareness of what I didn't know.

And then you say, "*Okay, now we understand some of the organic chemistry and the biology that goes along with this, but how do we actually get an electromagnetic signal that comes out of that? How does that work?*"

In other words, it's one thing to just tautologically define a photo-receptor as "the thing that tells the brain about light" and then say that our eyes are filled with photoreceptors—which is the picture I unthinkingly had before opening your book—but it's quite another to start breaking things down into these components so that you start appreciating—physiologically, biochemically—how things actually work. The result is that reading your book made me continually recognize that I had taken lots of this stuff for granted. Is that the sort of effect you were going for?

JG: Yes. When I was writing the book, I discovered things like that for myself. That particular process was familiar to me, but the processes that lead to that were not so familiar. For example, I had never thought about how essential it is that the photoreceptors be lining the back of the eye, as opposed to being somewhere else: that without that physical structure of the photoreceptors at the back and a small aperture at the front of the eye, image formation wouldn't be possible.

HB: I should also add that you often highlight many fascinating historical aspects of the story. You talk a fair amount of Kepler and Helmholtz, for example.

Did you learn some of this as you went along, or did you have a clear sense of that when you started? Personally, I found that to be really interesting. Kepler, for example, I'd always associated simply with the discovery of the elliptical nature of planetary orbits—which was a big deal, of course, and what he is most famous for, but it turns out that he did a heck of a lot more than that.

JG: Yes, he did a heck of a lot more. I didn't know any of that either before I began the book. I was very intrigued by the fact that Kepler did all this work on optics because he already knew about the elliptical orbits but realized that he was going to have a political problem pushing that view. He recognized that it was going to be controversial, and he essentially needed to get tenure. He needed to keep his job, and to do that he needed to produce something. So he did this optics work as kind of a side project to establish his credentials.

HB: So it was a deliberate, tactical initiative?

JG: So the historians say. It's often hard to know for sure what someone was thinking, but it seems that he made that initial discovery and then needed to justify his credentials. And he recognized that unless you understood what the eyes were telling you about where things were, you couldn't really be sure that the measurements you were making in astronomy were accurate.

HB: All his work on vision, then, was associated with some meta-level justification for these measurements of elliptical orbits, because, after all, your eyes could deceive you. So first he had to show how vision worked.

JG: Exactly.

Questions for Discussion:

1. Does this chapter make you better appreciate the different levels of understanding involved in "a scientific explanation"?

2. Do you think that, in general, popular science books written by younger scientists have a different perspective than those written by older ones?

III. Frames of Reference

Integrating sensory systems

HB: In *Making Space*, you talk specifically about three different systems—vision, touch and hearing—how they work and how they are integrated.

JG: Yes. The question that I've been really interested in centres on something that also has its roots in astronomy: frames of reference. Ptolemy's idea about how to think about the motion of planetary bodies was that the Earth is at the centre and everything rotates around it. Sometimes those paths are complicated and strange, but the Earth is the centre of it all.

Well, it turns out that if you treat the Sun as the centre, a lot of the orbits become much more regular and sensible: the Earth goes around the Sun just like all the other planets. In other words, the motion of the planets relative to the Earth is complicated, but the motion of the planets relative to the Sun is straightforward. That's an example of how choosing a different centre of your frame of reference can really change the way you're looking at the available information.

The problem for the brain is more local and specific. The visual system—these photoreceptors—are giving you a representation of where visual information is, where objects are in the world, and that frame of reference depends on where the stimuli are with respect to the array of photoreceptors. In other words, it depends on where the stimuli are with respect to your eyes.

Well, we can *move* our eyes; and we *do*. In fact, we move them *a lot*—about three times per second—and we move them *really fast*: at a speed of about 500 degrees per second. That means that, if you were to allow the eyes to continue moving, they'd be able to rotate around one-and-a-half times in the span of one second. They aren't able to move quite that far, of course, but that gives you a sense of their range and speed.

That's a lot of eye motion that the brain has to deal with, to compensate for. It has to assemble the snapshots that are taken by the photoreceptor array at each of the different positions that your eyes might be looking.

Now, that's just vision. If you then extend this problem to include some of the other senses, like the auditory system, it's important to first recognize that, of course, sounds aren't affected by how the eyes are moving. The auditory system is using a different frame of reference for figuring out where the sounds are located, which is based on subtle cues that are different across the two ears.

A sound that's located on one particular side will arrive in that ear first and will be slightly louder in that ear than the other. The brain has to compare the signals arriving in one ear with the signals arriving in the other to compute the angle that the sound is coming from.

HB: Which, as you point out, can often be very difficult. You give a concrete example that most of us are familiar with: trying to locate where the beep from a smoke detector is coming from, which is very hard to do.

JG: Yes. It's a frustrating problem for all of us to find a beeping smoke detector.

In general, then, the auditory system is computing sound location based on cues that are fundamentally anchored to the head, while the visual system is computing visual locations based on cues that are fundamentally anchored to the orientation of the eyes. So every time your eyes move, you're yanking your visual scene around to some new position with respect to your auditory scene.

I've been really interested in how the brain fixes that, how it puts those two signals into a common frame of reference so that you can do things like use lip-reading cues to help you understand what someone is saying. If you don't correctly associate the lips of the person who's speaking with the sound of that person speaking, then you can't make use of that supplementary information.

HB: That's interesting. I've certainly noticed that if I'm not actually watching somebody directly—if I'm just overhearing something, say—it's generally much harder to tell what is being said. Is that very common?

JG: I certainly find that. I think that's one of the reasons why I really like talking to people in person, or video calls, because then I get the lip-reading cues. I don't have any known hearing deficit, so I think this is something that is fairly common. We all certainly lose hearing as we age, and I think using the lip-reading cues is more and more important.

HB: Recognizing how rapidly our eyes are moving together with all the necessary adjustments that need to be made as we move our heads through space makes one start to appreciate the mind-boggling structure that must necessarily exist to compensate as we're doing the most normal things imaginable.

Once again, there's this sense of realizing just how complicated all of this really is. When you look at it from an individual system perspective, let alone integrating it with these other systems—which naturally makes it even more complicated—it's really remarkable.

JG: Yes, it's really amazing. It's both complicated and very logical. That's something that I was hoping to convey in the book: that this isn't really a mystical problem, this is something that we can reason about.

We can think about the physics and what the biological sensors are. We can look at it as a problem that the brain has to solve and then we can think about what kind of neural circuits might be accomplishing those tasks. I think we can make quite a bit of progress. It isn't all just hand-waving, "*...and then a miracle occurs*".

Questions for Discussion:

1. Why does the motion of our eyes present a problem in understanding how the brain puts together sensory information?

2. How, exactly, does the concept of moving from a geocentric to a heliocentric picture of the solar system help explain the "frame of reference" problem in coordinating sensory input that Jennifer is focused on?

IV. Mysterious Overlap

Fitting the pieces together

HB: One thing that struck me—and this is probably just a sign of my ignorance, but I'm guessing other people have similar levels of ignorance—is that you call some things neurons that aren't in the brain. You talk about sensory receptor neurons: cells that are detecting stuff and sending signals to the brain. That was something that took me aback a little bit. I thought, *Hang on, how can I have a neuron in my arm? That's not where it should be.* Is that a common misconception that people have—that neurons are only to be found in the brain?

JG: I don't know, actually. I don't think it's come up. But anything that's conveying information through electrical signalling, we call a neuron.

HB: OK. Let's talk a little more about touch now. What are the frames of reference involved there?

JG: The frame of reference problem for touch, and its integration with vision, is even more complicated because of the potential additional movements involved and the breadth of the relevant landscape of interaction.

With vision and hearing, you only have to worry about the motion of the eyes or ears with respect to the head. But when it comes to integrating touch into this—for example, if I hear a buzzing sound and then feel a bee stinging my hand—that could be happening in a number of different places. So the brain has to know what location on the body's surface was stung by the bee, and then it also has to know the posture of my wrist, my elbow, my shoulder, my head, and then the position of my eyes with respect to that. And that's just to be able to say that this image of the bee is coming from the same location that I'm feeling the sting.

HB: And that gives you a sense of how much processing power has to be involved in not only all of these systems individually, but integrating them together.

JG: Yes. And it happens fast too, because when you get stung by a bee, you have to deal with it immediately. You're not thinking it through.

HB: You start off the book with this joke about how 90% of neural activity is involved in figuring out where things are, and then you go on to say that you actually don't know that for a fact, but later on you explain why you think so much of the brain is involved in this and what the implications of that are.

The idea I took away from all of this is that these are very complicated systems, and from an evolutionary perspective, since so much of our mental activity goes into these systems, they must somehow do even more than solely locate things in space. Is that a fair way of looking at it?

JG: Yes, that's a very good way of looking at it. As you probably know, there is this popular myth out there that we only use 10% our brains. None of us actually know where that number comes from, so it's kind of a running joke in neuroscience to say, "*Alright, well, I'm just going to throw a number out there and say this is how much of your brain is involved with this, that, or the other thing.*"

But I'm half serious—or maybe a little bit more than half serious —because, when you look at it, there's an awful lot of the brain that's been identified as carrying some kind of information that's relevant to these kinds of processes that we're talking about.

There's a lot of the brain that responds to visual information, there's a lot that responds to sound, there's a lot that responds to touch, there's a lot that's involved in controlling movements— and movements are essential to understanding how to combine information across these different sensory systems.

If you were to say that all of those brain structures are really just doing those things—that it's their job to work on these spatial-processing, sensory, motor-control issues—there wouldn't be that much left for doing the things that concern, say, what makes us smart. How come most of the brain isn't involved in, say, language?

It turns out that, if you look at areas of the brain that seem to be involved in, say, language, or memory, or attention, or planning, or motivation, there's a lot of overlap between the structures that are implicated in those processes and the structures that are implicated in sensory and motor processing.

Questions for Discussion:

1. *Have you ever heard the statement that, We only use 10% of our brains? Will you be more sceptical of such claims in the future?*

2. *What do you think Jennifer means exactly, when she talks about the overlap between the structures implicated in sensory and motor processing and those responsible for actions like language and memory?*

V. Smell

An overlooked sense?

HB: I'd like to ask briefly about another sense that we haven't yet mentioned: the sense of smell.

As we're talking, I'm wondering, *Do I localize smell at all?* I think perhaps a little bit, but not very well, typically. Has there been much work done on that? Do we, in fact, localize smell?

JG: Humans are fairly poor at localizing smell, but smell is a very spatial sense for some other animals.

HB: Like dogs, I'm guessing.

JG: Yes. And I think they do it in a way that is both very interesting and parallels what we do in vision: constantly taking new samples as we move our eyes around and then stitching those samples together.

Similarly, I imagine that a dog sniffs in one location, then moves to some other location and sniffs there, and so on. It's by integrating and comparing those different samples that they're able to say, *"There's a gradient of smell. It's fainter over here, but it was stronger over there, so I'm going to keep going in that direction."* That's a process that you can see dogs doing, but it's one we don't have much personal experience implementing ourselves—mainly, I think, because we habituate to smell so rapidly.

The samples that we take on one side of the room versus the other side are so contaminated by habituation, by a decrement in our sensitivity to a smell after we've first experienced it, that it's very hard for us to make those comparisons.

HB: Are there people who compare and contrast the olfactory system of dogs and humans?

JG: I assume so, but I really don't know much about it. I'm just telling you what my gut impression is from watching dogs.

Questions for Discussion:

1. *Why do you think that smell is often so strongly linked to memory? What could be a possible neurophysiological mechanism to account for that?*

2. *Might there be more information in olfactory signals than we naively appreciate? How is it that some dogs seem able to "sniff out diseases"?*

VI. Brain Maps

Making a picture

HB: So there are all these complicated systems, and things get even more complicated when you think about how to integrate them. But there's also the question of how, exactly, these systems are represented in the brain, which brings us to the notion of brain maps. What are those all about?

JG: I'm going to start my answer to that by going back to my history of how I got into the problem. As I got thinking about how you might convert signals from one frame of reference to another, I quickly realized that I would have to grapple with the question of how the information is coded in the first place. In what way are neurons expressing information about where a stimulus is?

With the visual system, you've got this array of photoreceptors, and the location of activation within the photoreceptor array corresponds to the location of the stimulus in the world. That's a kind of map. You could also think of it as a photograph or snapshot, but we refer to it as a map. The activation of neurons in one location of the brain corresponds to a stimulus occurring in a particular location in the world.

In the auditory system, there was some information about how the neurons represented sound location, but it wasn't completely clear what the representations were. You can't infer it from first principles.

In the visual system, you can infer it from just simply knowing the structures of the eye and the optics and the way light travels— you can determine what's going to happen.

But as I said before, when it comes to the auditory system, figuring out where sounds are located is intrinsically a computation that the brain has to be performing by comparing stimuli from the two ears.

As I started trying to build models of how you might convert auditory signals from an original, head-based frame of reference into an eye-based frame of reference for communication with the visual system, I realized that how that computation might unfold would be different depending on whether there's a map for sound location or if sound location is represented in a different way, which we've come to call a meter.

What I mean by that is that, instead of knowing which neurons are active in signifying where the sound is coming from, you could also imagine focusing your attention on the amount of activity in a given population of neurons to signal whether or not the sound is more to one side or the other.

HB: The result would then be location-independent in the brain, presumably, in terms of those neurons—or at least not as dependent?

JG: That's right. Which neurons are active isn't so much the important factor in this case, rather, it's a matter of how vigorously they're responding. So you can imagine left-preferring and right-preferring neurons, and the ratio of activity between those two pools would tell you whether or not the sound is straight ahead—if the two pools are equally active then that means the sound is straight ahead. If the pool on the left is more active, that means the sound is coming from the left, and so on.

It appears that in humans, monkeys, and certain other mammals, that's exactly the kind of code that seems to emerge from these comparisons across the two ears.

That's an interesting result because it suggests a simpler way of converting auditory signals into an eye-centred frame of reference than would have been the case if it were a straightforward map.

And this, I think, exactly relates to what the brain has to do. In order for the brain to convert something from a head-based frame of reference to an eye-based frame of reference, it has to know where the eyes are, and then essentially subtract the eye-with-respect-to-head signal.

The mathematics of how you would accomplish this is really easy if you're representing both where the sound is and where the eyes are in a common, level-of-activity-based way. You can effectively subtract the signal that corresponds to the eye position from the signal that corresponds to the sound location with respect to the head, and what you're left with is the sound with respect to the eyes.

Questions for Discussion:

1. What does Jennifer mean by "a common, level-of-activity-based way"? What would be examples of real-world situations that would correspond to that scenario? What would be examples of real-world situations that don't?

2. Might other sensory input also use the notion of a "meter" that Jennifer describes in this chapter? How could that be tested?

VII. Ice Cream Cones and Multiplexing

Same neurons, different functions?

HB: But here's what confuses me whenever people start talking about measuring signals in neurons for this or that specific activity: that's not the only thing going on in my brain. I may be thinking about an ice cream cone, or I might be wondering what time it is, or what have you. How sure am I, if at all, that there's a one-to-one functional map between these neurons and what it is that you're focused on measuring? Each one of these neurons might wind up doing lots of different things at the same time, so it could get extremely complicated.

JG: That's a great question, and we've been really grappling with that lately. There are some experiments we've been doing to try to figure out whether our neurons may be switching back and forth between different roles.

An experiment that we're working on right now involves presenting two sounds at the same time. We're trying to tell whether or not the neurons are effectively multiplexing signals related to those two sounds. Do they alternate back and forth between coding one sound and then coding the other? Are they also switching back and forth between dealing with sound and maybe spending part of their time dealing with, say, auditory imagery, or even something that's more of a cognitive, thought-based role that might use the same neural infrastructure?

HB: And, of course, if they are changing activities somehow, then you presumably need a whole other level of explanation, a sort of switching mechanism, to understand how that happens. You need some sort of mini-central processing unit to structurally tell them when they should be doing all of these different things.

JG: That's right. But something like this might be part of the explanation for why you see so much overlap between the areas of the brain that are related to sensory and motor signals and those that are involved with attention, memory and language-related tasks.

It's possible that the mechanisms involve different neurons that are co-located, but it's also possible that it boils down to a common population of neurons; some of the time they're doing one thing, and some of the time they're doing something else.

HB: This is obviously second nature to you, but, again, it just emphasizes to me how incredibly complicated all of these processes are when you look carefully enough. There are so many things that I had completely taken for granted.

In *Making Space*, you talk about the importance of thinking about how you can actually build something, this idea that you can't really understand something until you can imagine how you'd go ahead and build it yourself. The more you start looking at this carefully and asking these sorts of questions—at least for me—the more you start realizing how incredibly difficult it would be, even just in principle, to construct these sorts of brain structures from scratch.

Questions for Discussion:

1. How likely is it that some neurons are singly dedicated to one task while others can multitask? Is it reasonable to believe that different types of neurons exist, or do you think that if one neuron can multitask, then they all can (and do)?

2. Will we ever be able to build an artificial brain from first principles? If so, how would it differ—if at all—from an artificial intelligence type of computer program? If not, why not?

VIII. Navigating Rats

Place fields and memory

HB: I'd like to return for a moment to this notion of brain maps and ask you to put in context for me the work of John O'Keefe and his colleagues, who won the 2014 Nobel Prize in Medicine or Physiology for their work on a "GPS system" for the brain. How does that relate to the sorts of things that you're telling me about?

JG: This takes us to a level of spatial processing that transcends what's in the "here and now"—it's more about how you move through the world and how you navigate and travel from one place to another. An area of the brain that seems to be involved in that is the hippocampus and some associated structures that we put under the heading of the hippocampal complex.

We knew, dating back to the 1950s, that the hippocampus is critically important for memory. That was first discovered based upon a famous patient called H.M., whose full name was Henry Molaison.

He had a bicycle accident as a child and had subsequently developed intractable epilepsy, for which he opted to have surgery. The surgeon removed the hippocampus on both sides, and H.M. was no longer able to form any new memories of episodes that had occurred in his life after that point. He did retain some kinds of memory, but most very simple events—like going to the grocery store—weren't registered. It was clear, based on his case, that the hippocampus was critically important for memory.

O'Keefe and his colleagues studied the hippocampus of rats. And what they discovered was that, if you measure the activity of neurons in their hippocampus, they seem to be responding selectively when the rat is in a particular location in the environment. Some neurons respond when the rat is here, other neurons respond when it's there—wherever it may be, there are some neurons in the hippocampus that are active when the rat is in that particular location.

This seemed really intriguing because, on the one hand, you've got lesion data in humans—replicated in animals—which suggests that if you eliminate the hippocampus you see a profound impairment in memory, while on the physiology side, if you look at what signals are present, the most obvious thing is that the neurons there are very sensitive to where you are in the environment.

There was subsequent work from May-Britt and Edvard Moser, who shared the 2014 Nobel Prize with O'Keefe, showing that there are other populations of neurons that seem to be carrying signals that relate to your progress as you move through the environment.

There are neurons that seem to signal that you've gone so many steps in a certain direction, and then another number of steps, and then another, giving you a sort of grid-like representation of how far you've gone.

HB: Did the rat have to be familiar with that particular environment? If you put the rat in a completely new environment, does it take a while for these neurons to start registering in the same way?

JG: If you put the rat in a completely new environment, the cells immediately have what are called place fields, but the place fields that they now have are unrelated to the ones they had in the previous environment where you tested them.

So if we had a rat here running around in this room, we could measure which two neurons were responding when the rat was in different parts of the room.

If we then take the rat into some other setting, both of those neurons would respond, but the relationship between the locations that each neuron is sensitive to would be totally random and different.

The neurons seem to be recruited for every setting, but there's some combinatoric process involved in representing what that environment is. It's kind of like the bits in a digital code: they're either on or off in all of the different settings, but the relative ordering might be totally different.

HB: I can understand if it's a familiar setting, but when you move a rat to a completely new setting where it's never been before, how does that process work?

JG: As best we can figure it, it seems that the representation corresponds to your sense of where you are. I'm anthropomorphizing here, imagining myself as a rat.

But if you put the rat in a room with only a few details or landmarks that give a sense of orientation, and if you then take the rat out of the room and rotate those landmarks—maybe disorient the rat by covering its cage with a towel while spinning it around, for example, and then bring it back into the room so it can't remember how it got there—what you'll find is that the representation of space in the hippocampus seems to shift to match those landmarks, as if the rat thought that the room was actually still in its original orientation. So you can kind of fool this representation of space.

Anyway, getting back to the link between memory and location in the hippocampus: you've got your lesion data telling you that there's a memory deficit, and your recording data emphasizing that there's a lot of information about location in the environment.

What I think about this—and I'm hardly the first person to draw this connection—is that your location in the world is one of the ways that you "file" relevant information for memory. For example, when I'm here in my office, I want to be able to call up memories related to the projects that my students are working on. There are specific memories that are relevant because I'm here in this place.

HB: We've all experienced that. If, for example, you go back to your old school where you haven't been for 25 years or so, all of a sudden memories will start flooding back to you. This is a very common experience that most people have had, at least if they're old enough to have experienced it.

JG: That's right. When you go to a reunion of some kind, things that you probably wouldn't be able to remember if you weren't back in that setting suddenly seem accessible to you. We don't have everything loaded into working memory all the time. We have to be able to click on that particular file and open it, so to speak.

HB: So there's a sense that "clicking on the file" has something to do with this spatial context, with knowing where you are in a particular place—that there's a direct link or, at least some link, between the two.

JG: Yes, exactly.

Questions for Discussion:

1. In this chapter, Jennifer speculates that "location in the world" might be one of the ways that we "file" relevant information for memory. What might some of the other ways be?

2. How might a "landmark" be objectively quantified? Doesn't this, in turn, depend on our sensory capacities? Might a dog regard certain smells as "landmarks"?

IX. Neuroplasticity

Phantom limbs, cochlear implants and feedback

HB: I want to explore the link between memory and space a little bit more, but before I do I'd like to ask you some questions about mental maps and this phenomenon of phantom limbs, which is something you also mention in *Making Space.*

I understand that there's a correspondence between some mental map and a neuron registering some sensation in a different place in my body, like my leg. But in some instances, if the leg is removed, people might still feel some pain—so-called "phantom limb pain".

As I understand it, my sense is that they still feel this pain because that map still exists at some level and sometimes this map somehow gets activated, so that even though the sensory receptors are no longer there—because the leg is gone —the corresponding map is still in place in the brain, and those neurons that are part of that mental map somehow spontaneously fire.

I have two questions linked to that. My first question is, 'spontaneously fire' means...that you just don't understand what's going on, right?

JG: Well, yes, in fact. That's a really good question. That's one of those things where, if you dig a little bit, you realize that what we've got is a fairly hand-wavy type of explanation.

The connections to any brain area come from a lot of different places, and only some of those connections are going to have been disrupted by the loss of the limb. You could imagine that maybe that spontaneous activity occurs because of the inputs that still exist that are coming from other places.

You could also imagine that it really *is* just spontaneous: that maybe sometimes, just by random chance, you get close enough to the threshold for producing a spike of electrical signals—that it just happens without there being any particular input.

HB: Well, I'm not really comfortable with the notion of things happening by random chance. To me, that seems like just another way of saying, "*It's spontaneous,*" or (as I like to put it), "*We don't know what's going on.*"

JG: Fair enough. I suppose the difference is whether or not it's spontaneous in the sense that it comes from within the neuron, or spontaneous in the sense that it comes from sources that we're not measuring.

HB: Okay. My second question concerns neuroplasticity. The assumption here seems to be that I have certain neurons dedicated to certain tasks. Granted, it could be much more complicated than that—because the same neuron, as we talked about earlier, might be involved in four different things, or ten different things. But, forgetting about that for the moment, the assumption is that a particular neuron is forever dedicated to some particular tasks.

But my understanding is that that's not actually the way it works either: neurons can change their focus and orientation. How does that get incorporated into all of this?

JG: Well, that really is a miracle; and it's a profound one. Yes, the way your brain is wired up is probably quite different from the way my brain is wired up, yet we're able to do some of the same things. You can walk, I can walk. You can talk, I can talk. We're able to do those things based on plasticity that refines these connections.

I think one of the things that has been really exciting in brain science is the discovery that we have even more plasticity than we previously appreciated, and that we have a tremendous ability to recover from deficits and from changes to our nervous system.

A great example of this is the cochlear implant. This is a device for people with hearing loss, which involves placing an array of electrodes in the cochlea, the auditory portion of the inner ear. Hair cells in the cochlea are the auditory equivalent of photoreceptors in the eye—they're responsible for transducing the physical vibrations that make sound waves into electrical signals. If you have hearing loss that impairs that process, if those cells have died or something, then you can replace their action by electrically stimulating the nerve fibres that they make contact with.

However, you can't do it in anything remotely close to the way that the original hair cells would have fired. There are thousands upon thousands of these hair cells and they're each making a unique set of connections. But it turns out that you can do pretty well by putting in about eight electrodes; and patients who have an electrode array consisting of those eight electrodes can do pretty well at starting to understand speech.

I've talked to a friend who got one of these devices, and her experience of it was that it gradually began to actually sound like it had originally sounded. She lost her hearing as an adult, and once she got this prosthetic device, over time her experience of sound became similar to what it had been when she was a child.

In my view, that has to be related to plasticity. That has to be the brain saying, "*Well, I'm going to take this bizarre input and re-route things to make it sound kind of like it used to sound.*"

HB: And there also must be some feedback going on, presumably, for the brain to adapt over time.

JG: That's right. And one of the interesting things is the question of how the brain actually gets that feedback. It gets that feedback from vision, because it has to be from something outside of the realm of hearing that tells you this. For example, my friend told me that one of the training strategies she was told to use was to read a book and listen to it on tape at the same time.

HB: So, empirically, people have realized this. The people who were telling her to do this, presumably, are not necessarily neuroscientists, but people familiar with what has worked for many others in her condition.

JG: Yes—that and some logic.

Questions for Discussion:

1. *Why do you think that Howard is so uncomfortable with the notion of "spontaneous firing"? Why does he make an equivalence between "randomness" and "not knowing what's going on"?*

2. *Why does Jennifer say that the feedback from a cochlear implant "has to be from something outside of the realm of hearing"?*

X. Evolutionary Mechanisms?

Repeat performance?

HB: People have known for a very long time about the connection between memory and space. As you know well, and as you cite explicitly in *Making Space*, the ancient Greeks had the mnemonic technique of a memory palace: this notion that we can actually remember many things in a reliable way and for a longer period of time if we train our minds to be able to put them in a spatial context, in a "memory palace", as it were.

There was a book that came out a while ago that emphasized this (together with other memory-related issues), called *Moonwalking with Einstein* written by Joshua Foer. The basic premise of the book is that Foer, who is a journalist, finds himself covering the "mental Olympics", which feature individuals who have incredibly prodigious memories. He assumes that they're just freaks of nature, but they tell him, *"No, in fact, anyone can learn this. If you want to, we can elevate you to the same level that we're at in a relatively short period of time."* He is naturally sceptical, but embarks upon this journey and, lo and behold, he actually *is* able to successfully complete this himself, bringing himself up to the same level by using a compilation of these techniques.

It's well known empirically that there is a very strong connection between spatial representations and memory. My understanding is that, to you, that's not only intriguing, but a sign that maybe there are equally deep, integrated links between, not just spatial representations and memory, but spatial representations and thought, writ large.

These systems, and their integration, are so complex and detailed, and require so much processing power, that that's suggestive of the fact that there might be a significant overlap between spatial processing and things that we might not logically, or immediately, assume have to do with spatial matters at all.

It may well be that most of the things we may be thinking about — or at least a good many of the things—are somehow, in a way that we might not immediately perceive, tied to the notion of spatial systems and their integration. Is that your view?

JG: Yes. This comes back to evolution and how the brain evolves. How did we get to be so smart compared to, say, jellyfish, or primordial ancestors who clearly did not have the cognitive abilities that we have?

A general problem in evolution is to envision how simple events, like the mutation of an individual gene, can produce an organism that functions better than the other organisms that don't have that mutation—because, usually, when you tweak something, you make it worse. Intermediate states in the course of evolution are often hard to envision.

For example, having hands with five fingers and opposable thumbs is very useful, whereas it would not be particularly useful to have a stump of a thumb. How do you end up getting the whole thumb?

One thing that may be happening is that modules in the brain might be duplicated through a fairly simple set of mutations so that you might take a structure that's working well, and maybe one small change means that you now have two of those. And if you now have two, and the one before was sufficient, you find yourself with an extra that can be used for something that you weren't originally doing.

HB: And so you're able to experiment, in a way, with that extra one without damaging the overall system.

JG: Yes. And because of plasticity, you can imagine shaping that extra one to be doing something quite different.

And the thought is that perhaps there's still some history that the duplicated module retains based on where it originally came from and that it would share some of the similar structures. The circuitry might look similar to what's present in the original area because perhaps it's getting some of the same kinds of inputs, but it wouldn't be doing the same exact things. It would be doing something similar, but on a different type of input.

And it may be that spatial processing originally arises as some thing that's essential for that first module to do, and that when it gets duplicated in that second module, you still have all this spatial infrastructure, only now you're going to use it to do things like think and reason about abstract concepts that might easily be equatable to something spatial, but aren't, in and of themselves, spatial.

I'm essentially elaborating on an idea developed by George Lakoff and Mark Johnson, who noticed that there's a pattern to the metaphors we use in language. For example, we talk about social relationships using spatial language. We say that you're "close to your parents" or you have a "distant relative". To call those examples metaphors is stretching the definition of that word, but it's nevertheless a sort of metaphorical use of terminology that has a spatial meaning.

HB: That's suggestive, but then you cite some examples of people who, as I understand it, are actually studying these linguistic metaphors using fMRI and other brain-imaging devices. And it turns out that indeed there's a statistical correspondence between these metaphors that they're using and the relevant motor or sensory systems in the brain.

JG: Yes. There are just a few of these studies out there, but they're really intriguing.

For example, if you give subjects words that relate to actions that involve a particular part of the body—'kick' involves your feet, 'lick' involves your mouth, and so forth—you see a pattern of brain activation in the region of somatosensory and motor cortex that roughly seems to correspond to the layout of the body map in those structures.

So the activation that you see when you present words like 'kick' tends to be more towards where the legs are, while words like 'lick' tend to be more towards where the face is. It's not a perfect correspondence, and these things are never as straightforward as one might like, but it's a really fascinating observation that perhaps when you're thinking about something like that, what you might actually be doing is simulating some of the sensory and motor attributes that are related to that concept.

HB: OK, so not perfect, but there is still a clear correlation.

JG: It's clear that the areas of activation are different, and that they're laid out in the order that the body is laid out. It's a little less clear whether they exactly map on to the particular locus in the body.

HB: But a priori, if I know nothing about this, it's not surprising that there wouldn't be an exact match, because these neurons are doing all of these different things at the same time. So I wouldn't expect that there would be an exact match anyway.

JG: Yes. To demand that there be an exact match would be putting too strict a criterion on it.

HB: Because then the neurons in the foot would be doing nothing else but waiting around for me to think of 'kick' and so forth. But they're doing all sorts of things in the meantime.

JG: That's right. And if they were doing that, then how could you be responding to sensory and motor input? So there has to be some difference between the circuits that are responding to what's coming in and the circuits that are thinking about things.

Questions for Discussion:

1. Do all languages use spatial metaphors to roughly the same extent? If not, what, if anything, does this imply for this hypothesis? Might it be possible to probe matters further by explicitly contrasting different language speakers?

2. How are the issues in this chapter related to our understanding of how just thinking about an activity (say, playing tennis) can stimulate motor areas in the brain linked to that activity?

XI. The Road Ahead

Testing neurons for contrast

HB: How might you test this exciting hypothesis? What sorts of experiments do you have in mind, and what experiments would you think about doing if you had infinite time and infinite money to be able to probe this a little bit more?

JG: That's a great question. I'm still in the brainstorming phase of thinking about this, but one of the things that I've been thinking about is that one of the central, organizing principles that we see in the visual system is its tendency to draw contrasts to emphasize where things are changing. There are neurons that start to do this as early as the retina, before signals have even made their way into the brain. So I've been wondering if there is conceptual or cognitive information that would benefit from using that kind of infrastructure.

We draw contrasts all the time. For example, words can be opposite each other in meaning; we can use negation to change the meaning in the opposite direction: good versus bad, running versus not running. Those contrasts exist in the domain of meaning, and I wonder if the neural infrastructure to do that involves something similar to what's done to emphasize contrast in the domain of visual information.

If I had a way to explicitly test that, I'd be doing it. Maybe somebody reading this will have a great idea for how to test that.

HB: And which diagnostic tools do you think might be best to examine this in humans?

JG: In human patients who are undergoing surgery, sometimes recording electrode arrays are implanted to prevent disastrous scenarios, like what happened to H.M., so that there's more information about what the particular brain structures that are in question for removal are doing.

You can measure the activity of individual neurons, or small groups of neurons, in human patients, and you can get information that has a high temporal resolution. That's pretty well the gold standard for measuring what the brain is doing.

There are a few groups that have been looking at the activity patterns of neurons in various areas, looking at how selective they are for different kinds of images. Some neurons are sensitive to people, as categories. For example, there are neurons that have been identified in some subjects that are responsive to the actress Jennifer Aniston and only images of Jennifer Aniston—but *any* image of Jennifer Aniston, regardless of lighting, regardless of what she's wearing, regardless of what angle, and so forth.

One of the things that intrigued me about this—and this was probably just a short line in the paper that presented this finding—is that the Jennifer Aniston neuron that they found only responded to Jennifer Aniston if Brad Pitt was not also in the picture. This seemed to me an indication of a possible contrast, a strange form of contrast, but a type of contrast nonetheless.

There is always the risk of collecting stamps that aren't indicative of any larger pattern, but I think it's often worth thinking about things before we have data, before we have any idea what is going to turn out to be the overall pattern.

HB: Absolutely. Any last questions? Anything I missed that you'd like to talk about?

JG: I can't think of anything. It seems like we've pretty much covered everything.

HB: Thanks so much, Jennifer. I've really enjoyed talking to you.

JG: Thank you very much. I really enjoyed it too.

Questions for Discussion:

1. Do you agree that the lack of response of "Jennifer Aniston neurons" to pictures of both her and Brad Pitt represents a "form of contrast"?

2. Will increasingly accurate brain-imaging technology eventually render traditional "behavioural psychology" obsolete?

Continuing the Conversation

Readers are encouraged to read Jennifer's book, *Making Space: How the Brain Knows Where Things Are*, which goes into considerable additional detail about many of the issues discussed during this conversation.

Vision and Perception

A conversation with Kalanit Grill-Spector

Introduction

Facing Facts

As for most phenomena, the Ancient Greeks had a theory of vision, sometimes called "emission theory". The idea, often attributed to Empedocles, was that light shone out of the human eye and lit up objects in our visual field, making perception possible.

There were, of course, significant problems with this theory. If everything came from our eyes, why might we not be able to see equally well at night? This led Empedocles to postulate some relationship between these "eye rays" and those from other sources, such as the sun.

A few hundred years later, Euclid pointed out that, according to this view it was difficult to understand how, by closing and opening one's eyes under the night sky, we might suddenly be able to see the stars, which were presumably a long way away.

So, like any scientific theory, there were a few outstanding issues. But nevertheless, the "eye ray" theory of vision held sway for centuries. Serious doubts only began to arise through the work of the 10th century Arabian physicist Ibn al-Haytham, who promulgated a competing "intromission theory", which states that our visual perception of some external object is stimulated by something emanating from the object itself (in this case, light rays).

But whichever way you look at it, a key question remains. Once the core visual information has reached us, how do we process it? Just because light rays from a face, say, reach our optic nerve, that doesn't necessarily guarantee that we will appropriately visualize the face. And it certainly doesn't guarantee that we will recognize it.

This question is now duly recognized as a pre-eminent one. But it certainly wasn't always appreciated as such, even in recent times.

Kalanit Grill-Spector, the Principal Investigator of Stanford's Vision & Perception Neuroscience Lab, smiling, told me that only forty years ago, MIT computer scientists were so convinced that vision was simply reducible to sufficiently powerful computational algorithms, they assigned the question of how it works to some unfortunate undergraduate as a summer research project.

The reason that this was a laughable idea (for everyone other than the poor student, who might well have been unable to appreciate the humour in the situation), is that it turns out that the neural processes that underlie vision are so sophisticated and well-developed, current estimates are that they take up some 30% of all brain function. And this massive neuronal effort, ironically, is why we have taken so much of vision's complexity for granted for so long.

> *"The reason why vision research is so interesting," Kalanit enthusiastically related to me "is that an awful lot of the brain is involved in doing vision. And the reason it looks effortless is that there's a lot of machinery that is working away without us consciously having to activate it."*

Kalanit specializes in facial recognition. Much of the impetus for her current work began with a discovery of a particular region of the brain along a specific crease or sulcus—she obligingly puts it into non-technical terms by calling it a "dimple"—that seemed remarkably similar across a broad sample of the population. In other words, virtually everyone seemed to possess this special dimple, but it had hitherto gone unnoticed by the anatomical textbooks until she and her postdoc, Kevin Weiner, discovered it.

Having found the dimple, the next logical step was to investigate what particular sort of neural processing it might be involved in.

> *"It turns out that there is some dedicated hardware in your brain that seems to be involved in processing faces. This is a discovery made*

by Nancy Kanwisher in 1997 using fMRI. Initially, she thought that there was just one area, a module in the brain—she described it as a blueberry-sized module—that just processes faces.

"What we've found over time, from 1997 to today, is that there are, in fact, multiple regions that are organized in a very systematic way. And we've recently found that they're very predictable anatomically— they happen in the same part of every person's brain.

"We've started another collaboration with people who look at the histologies—the anatomical make-up of the brain. You can't do this sort of work with a living subject, because you have to look at slices of the brain under the microscope and you therefore need post-mortem brains. But they have discovered different regions of the brain that clearly seem to have different hardware.

"But we naturally can't test this hardware directly by mapping function to their data, because they're using post-mortem brains. But it turns out that the location where they've found these sorts of specialized hardware correspond well with a unique anatomical region on the sulcus that we'd recently found—a sort of dimple.

"So this led us to think that perhaps this special hardware in this region might be linked to some specific processing that's relevant to our perception."

Exciting stuff. But how on earth might it conceivably be tested? After all, many of the key results in our understanding of these issues necessarily seem to rely on either non-invasive measurements of subjects using fMRI (where no direct testing is possible) or anatomical analysis of post-mortem brains (where feedback is obviously impossible).

But sometimes, you can just get lucky.

"Once in a while, we have the opportunity to directly record from the surface of the brain. And this happens for subjects who get evaluated for surgery for epilepsy. There's an epilepsy clinic here at Stanford that treats patients who have intractable epilepsy that doesn't respond to medication. The doctors bring them in for testing to try

to evaluate where the seizure starts to determine whether or not they might safely be able to perform highly-localized surgery on them.

"They come to Stanford for a week or so and have electrodes implanted on the surface of the brain, waiting for a seizure to occur so that the doctors can track it. Sometimes, if the patients are willing to help us, we get to work with them during this waiting period to do additional tests for our research.

"In one particular case, it so happened that the doctor implanted an electrode in exactly the same part of the brain that we were looking at, and we were able to run a small current through the electrode and test things directly. Let me show you what happened..."

Empedocles, needless to say, would have been amazed. And so will you.

The Conversation

I. Neuroimaging

A transformative technology

HB: Since you kindly gave me a tour of your fMRI facilities here at Stanford, let me just dive in and ask some questions about that. How does functional MRI differ from the normal MRI machine that most people are familiar with—for medical injuries and so forth—and how does it differ from PET scans which people also might have heard of?

KGS: fMRI has really been a revolution in cognitive neuroscience. It started in the early 90s. It's the same machine that does the MRI for a knee scan, for example. The only thing is that, instead of measuring the tissue, we're measuring changes in brain metabolism.

When you use your brain, to do some sensory processing for example, your brain uses oxygen, and changes in oxygenation levels affect the local magnetic field; this is what's picked up by the scanner. The patient is placed inside the scanner, his or her head is placed in a coil, and this coil picks up signals in the brain that are linked to neural activation in the regions that are activated by whatever task the person is doing.

HB: When I first heard about fMRI I was confused about that because I thought, *Oh, they're measuring the brain, so they must be measuring the electrical signals*. Of course what they're doing is measuring, just as you said, the oxygen related to the blood supply that's flowing into specific brain areas because of neural activity.

KGS: Yes. We're not measuring direct neural activity. We're measuring a BOLD signal, a blood-oxygen-level-dependent signal. In fact, you might think there would be less oxygen because you've used it for the metabolism, but the brain overcompensates so you get an overflow of oxygenated haemoglobin. Really what we are picking up with a scanner is the amount of deoxygenated haemoglobin, and that actually gets washed off; this is why the signal goes up. So it's an indirect measure of brain activity.

The reason that it's different from PET—positron emission tomography—is that it's non-invasive. For PET you need to inject subjects with a radioactive material. That's an invasive procedure. With fMRI you don't inject anything. This has really been the power of fMRI technology because you can study the same person and run the experiments over and over again, or over time, or over development, or over the lifespan. This has been a really big breakthrough because you can peer into people's brains without doing anything to them invasively.

HB: And this was developed in the early 90s?

KGS: The first fMRI papers were published in *Proceedings of the National Academy of Science of the USA* in 1992. There were two groups that did this in parallel. One was a group at Bell Labs led by Seiji Ogawa. The other group was at Massachusetts General Hospital and the first author on that paper was Ken Kwong. Basically they conducted the first experiments where they showed people pictures versus no pictures—they had flashing checkerboards—and they could see an increase in the back of the brain where the visual cortex is located when people saw stuff versus when they didn't.

HB: So in your own studies, subjects go into this machine and you have them focused on doing particular tasks and thinking about particular things, or seeing particular things. How long do they stay in there for, in general?

KGS: They usually stay between an hour and two. Usually what we'll do is put the subject in the scanner and first run a brain anatomy scan. The reason we want to see their brain anatomy is because we are interested in which part of their brain is involved in what function.

This allows us to create—I'll show you later—these beautiful cortical reconstructions in which we can see the brain from all three dimensions, because the way we acquire information in the scanner is in slices, like in a CT scan, so we can get the 3D reconstruction. That takes about five to ten minutes. That's called just MRI, anatomical MRI.

HB: Let me just stop you there for a moment. Does this mean that people's brain anatomy differs significantly? I would have thought that this would be relatively constant. But it's not?

KGS: There are two things to consider here. The first is that, as a field, we are interested in how brain anatomy does change. I, for example, am looking at how it might change from childhood to adulthood. There are some things that will happen as a result of certain diseases, like Alzheimer's disease for example, where there are actually changes to the brain anatomy because of the disease.

The second point is that we want to look really closely at how function is implemented in each person's brain. So your brain and my brain have the same general pattern of what we call cortical folding—there are hills and valleys—but there are idiosyncrasies for each brain and we really want to understand the relationship between function and anatomy in each person's brain. So we want to take a detailed picture of every subject's brain.

HB: So there is really significant variation between different people? That's fascinating. I never would have thought that.

KGS: On one hand there is variety; on the other hand there is stability. One of the things we're trying to figure out is what is stable and what is variable across people. There is more variety than, for example, the hand; there are always five fingers on your hand. There is a little bit more variety in the number of cortical folds, but the big ones are very stable across people.

HB: Right—but I had interrupted you. You were telling me that these anatomical scans were the first things that you do.

KGS: That's the first thing we do and that takes between five to ten minutes. Sometimes we do another kind of scan called diffusion tensor imaging or diffusion weighted imaging. That lets us measure how water diffuses in the brain. We do this to look at the wiring of the brain—which part of the brain is connected to another part of the brain. There are these really big white matter bundles; they're called fascicles. Because they are myelinated, they are very directional, so the water doesn't diffuse in all directions.

HB: Hold on a sec. Where is the water in my brain?

KGS: Your brain is all water.

HB: That's not just my brain presumably?

KGS: Actually, all we are measuring in fMRI is how the magnetic field affects the water molecules. We're measuring hydrogen atoms basically. Because the water doesn't propagate freely, most of the direction of diffusion will be parallel to the fascicle, and we can measure the connectivity, the white matter connection, between one part of the brain and another part of the brain. This is another type of anatomical scan. This is a much more novel method in the field and the hope is that it will give us a wiring diagram of the brain.

HB: What are these fascicles?

KGS: The building block of the brain is the neuron. A neuron has a cell body, it has dendrites where it gets information, and it has axons where it transmits information. In a given brain area there are local connections, but some of the connections are long-range, and the long-range connections are like wiring fibres in your house. Suppose the back of the brain, the visual cortex, wants to connect to the front of the brain, the part that makes decisions. A lot of these axons come together in a bundle, and the bundle transmits information, let's say, from one lobe to another lobe of the brain. To get the transmission more efficient you want to myelinate it. It's kind of like…

HB: Like a cable.

KGS: Yes, it's like a cable. Sorry, all my electrical engineering terms are in Hebrew.

It's long range and you don't want the transmission to get broken, so that's why you have myelin. These are really a bunch of axons that are together in a big cable.

HB: For strength, from an evolutionary perspective?

KGS: The signal decays. Basically the electrical signal decays over distance, so in order for it not to decay you protect it with myelin.

HB: OK, so the water goes along these fascicles and when we look closely at this we can then see where these cables are and the connections. That's really cool.

KGS: So to summarize: first you get an anatomy picture of the brain. Then—sometimes at least—you run a diffusion scan. Usually you don't want to see these flow effects because if you want to measure the tissue and there is flow in the tissue, that's bad; you're going to get a distorted image. But if you actually want to measure how the water propagates, you get the water to diffuse in many directions. Let's say a hundred directions. Then you can build out of this what we call a tensor; this summarizes the average diffusion in each voxel. A voxel is what we get when we divide the brain into little cubes.

HB: These are like pixels?

KGS: Yes. A voxel stands for volume-pixel.

Basically we are trying to determine the main direction of diffusivity. We model this with a tensor, which is kind of like a cigar shape. It tells you which direction it goes and how directional it is.

This can be difficult because, although in some parts of the brain all the cables run the same way, there are places where the cables will cross each other and then that becomes more complicated to measure. What some people have been working on is how to model these crossing fibres so we can get a good wiring diagram of the brain.

HB: So you do the first scan. Then you may also do this diffusion tensor business. And then?

KGS: Then what most of us do in cognitive neuroscience is called functional magnetic resonance imaging. Basically you put a subject in the scanner, usually under a task.

When I study vision, I put them in the scanner, the room is dark and there is no visual stimulation. They have a screen in front of them. They see it through a mirror usually. Sometimes I don't show them anything—that's going to be my baseline. Then at certain intervals I show them different kinds of pictures. I want to see which part of the brain reacts to different kinds of pictures.

Basically what I'm looking at is local brain activity in response to the picture versus when I don't show any visual stimulation. This is where you get neural activation and this is what I measure with functional MRI. You asked me before, "*How long does this take?*"

HB: Yes.

KGS: It's up to the experimenter. When we scan little kids, we will put them in the scanner for fifteen minutes. If we scan people who can stay longer and concentrate for longer, we can scan them for an hour.

HB: Little kids can stay there for fifteen minutes in the dark?

KGS: Yes. We train them with a mock-scanner. It looks like a scanner, sounds like a scanner, but it doesn't take pictures of the brain. We have a motion detector that attaches to the forehead and we have a target that we use to train them to stay still.

Two things are really critical for us in an fMRI scan: first, that the subject is awake and that they do the task; and second, that they don't move because what we're really measuring is millimetres of brain. We can measure the resolution of fMRI anywhere between one to three millimetres. So you really need to be very comfortable when you're in the machine and not move your head because then we can't track the activity in that location over time.

HB: Does the head get set in place somehow? I understand that you have to hold it still, but is it aided at all?

KGS: No. We have the subject lie in the coil very comfortably. We will put some foam padding around them so it will be comfortable and feel snug. But it's really up to the subject to stay still. We really rely heavily on the cooperation of the subjects. We don't restrain them in any particular way.

HB: Is there anybody who couldn't handle it, who had to get out?

KGS: There are three things that might be annoying about the scans. First, the noise. It makes a loud noise so we give people earplugs, so the noise isn't too bothersome.

Second, some people don't like tight places. If they are claustrophobic they might not like to be in there. I've had one instance of a claustrophobic subject.

Third, occasionally subjects will get what's called peripheral nerve stimulation. The magnet in the scanner changes the magnetic field rather rapidly and some people, if they are muscular for example, might get some peripheral nerve stimulation; they might get some twitching. It doesn't happen that often, but it might happen. If the subject is uncomfortable we always take them out. We give them a squeeze ball because it's noisy; if they talk you might not hear them. They just press it and at any time we can stop the experiment and take the subject out. If they want to get out, they get out.

HB: Where do you get your subjects from? Where do they come from generally?

KGS: They're all volunteers. We get them from the local Stanford community. My lab also studies development, so we've scanned children and adolescents as well. We get them from the local communities through ads in Community Kiosk and things like that.

HB: Do you get people who are addicted to this, who just keep coming back again and again, unable to get enough of these scans, saying, *"Let me into the fMRI machine!"*

KGS: We have some screening. We don't want people to come for any ulterior motives. We need to filter them out. We tell people up front that we cannot tell them if they have any disease. We are doing science. If people want to make sure their brain is all right, we filter them out. They really need to volunteer because they are curious about research. They cannot have some independent motive, to get something diagnosed or what have you.

HB: Do you ever wonder, in terms of your sample size, that if you're pulling subjects from the Stanford region, that you're actually getting some sort of statistically skewed sample?

KGS: We have a variety of people. For studying vision I don't think that really matters as much, honestly. What we really care about is that people have corrected to normal vision. If you need glasses, you come with your contact lenses.

Over the years we have accumulated hundreds and hundreds of brains. We had a recent collaboration with a group in Jülich, in Germany, which looked at post-mortem brains, and the brains look the same. So in terms of the anatomy, and in terms of the field that I'm studying, I'm not really very concerned abut that.

HB: You mentioned that people started using these scanners to study vision in the 90s. I'm guessing that the technology has improved considerably since that time.

KGS: The scanners themselves were available before the 90s. The first imaging stuff in humans started in the 90s. The earliest scanners were 0.5 tesla, which wasn't strong enough to pick up signals from the brain.

The earliest neuroimaging scanners—what they used in the 90s—were 1.5-tesla scanners. Today the typical scanner is 3-tesla, which means it's sixty thousand times the magnetic field of earth, so it's really a big magnetic field. The stationary magnetic field does nothing to you—you don't feel it, because what we're really doing is making local variations in these magnetic fields—that's what we're measuring with MRI.

Here at Stanford, we also have a high-field scanner: 7-tesla scanner. There are even stronger ones around—there are some in Minnesota that are, I think, 14 point something, but they're often harder to use because they're not plug-and-play. Today the standard in the field is the 3-tesla scanner, because it's plug-and-play.

HB: So by plug-and-play do you mean you don't need advanced training and it's easier to set up and so forth? What do you mean exactly?

KGS: Yes. With a 7-tesla scanner, the physics is more complicated, so they are not clinically used. Getting the brain to be homogenous and stuff like that is more complicated and you need a physics team to run the scanner. That's what I mean when I say it's plug-and-play.

HB: Can you tell me why? I don't understand that.

KGS: Suppose you want to take a picture of the brain. You have different tissue types: the grey matter and the white matter. Grey matter is where neurons are. White matter is where the connections are. You want the intensity of the photograph to be equal across all the grey matter in the brain and you want all the white to be the same colour, the same intensity. You do this by having a homogenous field.

The external magnetic field needs to be very stable across the size of the object that you measure and you need to have really good coils to pick up these pictures. What happens as the scanner goes up, is that keeping the magnetic field homogenous over the size of an object like the brain becomes more difficult.

If you were scanning a mouse, which has a tiny brain, it's not difficult, because you use a really small bore. The physical challenge is to get the external field to be homogenous across the extent of the object you want to measure. That's why you need a physics team.

More generally, there are three things that have changed: the first is the intensity of the scanner as I've just been saying; the second is the coils that we use to pick up the signals; and the third is the resolution that we're measuring.

That has really been an improvement in terms of the sequences that we've been using to acquire these brain images, together with the coils. For example, my lab has moved into scanning with really small voxels. The standard in the field would be 3 to 4 millimetres, but we've been able to scan at 1.5 millimetres. It's a big change in the resolution.

Questions for Discussion:

1. Why do you think that Howard was so surprised at learning that there were measurable anatomical differences between the brains of different people? Do you find that surprising as well?

2. Might there be some studies of fMRI in cognitive science where ensuring that you get a very diverse sample of subjects is particularly important?

3. Why do you think that the earth's magnetic field is irrelevant to MRI measurements?

4. What sort of research questions do you think would be better suited towards using diffusion tensor imaging than solely functional magnetic resonance imaging?

II. Discovering Her Passion

A glimpse of the joy of vision

HB: Tell me more about your personal history and how you got into this field. How did that all start for you?

KGS: Well, I took the long route. I started with electrical engineering. I was trained as an electrical engineer, but I found it extremely boring.

HB: When did you start finding it boring?

KGS: Probably in my first year.

HB: Wow. Bad decision.

KGS: Yes. Anyway, there are aspects of electrical engineering that I really like. I really like the computer science part. So it's not true that everything was boring. But the hardware part wasn't very interesting for me. The problem for me with engineering was that I was a "why'" kind of person, and in engineering they don't care about answering "why"—they just want to make it work. It's not that engineering is inherently boring, it just didn't fit what I wanted to do with myself. I have good friends who are engineers.

Anyway, I was trying to figure out what I wanted to do with myself when I read a paper in *Scientific American* by Semir Zeki. He was one of the people who discovered the visual areas and what they do. He discovered the visual area that processes visual motion and another area that might be involved in colour processing in terms of a very intricate structure in the brain. And that made me think, *I want to do that*. So I started looking around to see where I could do that in Israel, where I was living.

HB: How far along were you in your undergraduate degree at the time?

KGS: I finished my undergrad.

HB: OK, so despite finding it pretty boring in your first year you persevered and finished anyway.

KGS: Yes. I had finished my undergraduate and was working, trying to figure out what I wanted to do with myself. Then I started taking some classes at the Weizmann Institute of Science. You could just come and audit classes.

At that time there was a major discovery in the field by Keiji Tanaka who discovered these columns for processing object features.

HB: What are these columns, exactly?

KGS: Basically the brain is organized so that neurons that process similar features are organized together. David Hubel and Torsten Wiesel got a Nobel Prize in 1981 for discovering the organization of the first primary visual cortex, called V1.

They found that it has this very nice organization in which neurons that process similar features—like orientation for example—are organized in this columnar way such that if you go perpendicular to the surface of the brain, they all have the same preference for orientation, but if you go parallel to it you get a slowly changing preference for orientation.

And one of the things that is still a question in the field is, *To what extent is this reflective of a general principle of organization?*

The visual system is organized in a hierarchy, and the higher order areas of this hierarchy were pretty opaque to scientists. One of the discoveries from Keiji Tanaka's lab was that these higher order areas also have the same kind of principles of organization.

So I went to this seminar and I didn't really understand anything, but everybody around me was getting really excited about that. I got stung by their excitement and I took these classes from Shimon

Ullman who was doing computational vision, but a lot of his computational vision was inspired by brain and cognitive science.

So I started going to these classes and it became clear to me that this is what I wanted to do. So I did a Masters in computer vision and then I actually did a PhD in neuroscience and computer vision.

My PhD advisor was Rafi Malach. While I was doing my Masters he did a sabbatical at Massachusetts General Hospital, and he came back saying, *"There is this new thing happening. We have to do it!"*

There was just no MRI in Israel at the time, so Rafi and I started doing that.

Questions for Discussion:

1. What percentage of people do you think enter university without a proper awareness of whether their course of study matches their personal temperament?

2. In what ways does Kalanit's experience reinforce the societal value provided by encouraging young people to freely audit courses from top-level universities and institutes?

3. To what extent do you think that Rafi Malach's experience argues for the importance of sabbatical visits to spur global research? Do you think that other areas of society—such as teaching or medicine—would also benefit from practitioners having regular sabbaticals? How might that be encouraged?

III. Vision Unveiled

Our current understanding

HB: OK, so let's talk in more detail now about what we know about the process of vision. I can imagine someone reading this who has a rough idea that when he sees something it corresponds to something going on in his brain—and perhaps he's heard of the optic nerve or the occipital lobe or something like that—but most people probably don't know—or maybe even care—much more than that.

A moment ago you mentioned V1 and other structures arranged in a hierarchy. Tell me more details about all of this—give me some overall sense of what our current understanding of vision is.

KGS: Vision is intriguing because it is very effortless for us. For example, we've never met before, but if you see me tomorrow you will most likely recognize me. People can recognize people from pictures within tenths of a second. Do you have any idea how many visual areas you have in your brain?

HB: I thought *I* was supposed to be the one asking the questions, but OK: I'll tell you the little that I know from my recent reading of things.

The brain is divided into four main lobes, four main regions. So I've got my frontal lobe at the front of my brain. I've got my parietal lobe up on top somewhere. I've got my occipital lobe at the back of my brain. And I've got my temporal lobes on the sides of my brain.

And my understanding is that visual information comes through my optic nerve to the occipital lobe. Then, my sense is that there are a couple of different processing streams of information. One is the so-called "dorsal stream", which has to do with "how" and "where" type of stuff—spatial location, planning, all the rest of that.

The second stream, called the "ventral stream" comes down into my temporal lobes, and is more the "what it is" stuff, that is tied to memory, recollection, and all that.

KGS: Very good. You would have passed that question in my *Introduction to Perception class*. You did well.

The reason I'm asking is not to put you on the spot—

HB: Too late now.

KGS: If I asked you, *"How do you solve a mathematical problem?"* You might think, *Well, clearly you need a brain, because it's difficult. You need a lot of brain power. It's uniquely human.*

It turns out that vision, while it seems very effortless, is hardly a simple process: it requires *a lot* of machinery. In the monkey brain about 50% of the brain—like a macaque's brain, for example—is devoted to visual processing. About 30% of the human brain is devoted to visual processing. But the way it's organized matters too, because it's not like a whole chunk is just devoted to vision. It's organized into smaller components that we call areas. There are specific criteria for defining an area.

I just want to give you a sense of this because you asked, *"How does vision work in the brain and why might it be interesting?"* It's interesting because a lot of the brain is involved in doing vision, and the reason it looks effortless is because there is a lot of machinery at work without you consciously having to activate it.

In the 70s there were major advances in computing, and when they formed the AI lab at MIT people said, *"Now that we have computers, vision should be trivial. We're going to solve it very fast because now we have the computer power."* So—this is a true story—they gave it as a summer project to one of their summer students. Suffice it to say that it hasn't been solved yet, and that was 40 years ago.

HB: That sounds to me like almost a classic definition of hubris.

KGS: I wasn't sure if this was an urban legend or a true story, but it turns out that it's a true story. There's an AI memo on it in the MIT archives.

Anyway, for me this is what makes vision so interesting, because it is a very concrete system: the goal is to see. Then within seeing there are concrete goals that you might want to figure out.

For example, recognizing people, that's the *what* stream or ventral stream you mentioned earlier; figuring out *where* things are, that's the dorsal stream; figuring out what the actions are, or integrating with the other systems of the brain, like multi-sensory processing, that's the third stream, the lateral stream.

Basically our goal is to understand how information gets from the eye, goes through the optic nerve to the occipital lobe, and then through a mysterious sequence of processing you get this *"Aha! moment"*. This is what we are trying to figure out, this sequence of processing.

HB: OK, I'd like to ask you more about that, but let's first go back to these streams. When I read about these different processing streams I thought to myself, *What's going on there?* First of all, why is it that all this information is being processed in different geographical areas of the brain? That seems odd. If I were God and I wanted to build a brain, why would I necessarily build it that way? That sounds like a curious thing to do.

If I want to, say, navigate through a room, I have to recognize objects—I have to say to myself, *"That's a chair"*—I have to rely on my memory, and at the same time I also have to be able to plan my route around them, I have to be able to avoid objects.

Yet according to this picture, those are two completely different processing streams. So that strikes me as a tremendously complicated structure. Maybe it could go all wrong. One stream could do this, one stream could do that. It doesn't seem like the best structure, evolutionarily speaking.

KGS: You ask a very deep and important question. There are two underlying hypotheses about why there is more than one area to begin with, and why we have these parallel streams. Why not just stick with one area that does vision? One way to think about it is from a systems engineering approach. Suppose you want to build an optimal system, what do you want to have it do? First of all, you want it to be rather fast because you have to react to a quickly changing environment. So you have to make the system very efficient in terms of its processing speed.

Second of all, you want to make it very robust. This is what people didn't realize in the 70s. They didn't realize that vision was difficult. *Why* is it difficult? Well, if I see you now and then I see you five seconds later, something has changed. You've moved your head. The lighting has changed. I'm in a different pose relative to you. So I'm never going to get the same image on my retina, the same photograph of you.

HB: So you have to abstract somehow. You have to integrate, use your memory, do all this other stuff.

KGS: Yes, that's one aspect. But another is that the world is 3D, but the pictures that we get are actually 2D. In fact we get two pictures—we have two eyes—and we can integrate this information to get a 3D sense of the world.

So a visual system not only needs to be *fast*, it needs to be *robust*— it needs to be robust with all this variation in the visual input.

The idea of parallel streams is that if you take a big and complicated problem and break it up into smaller problems so that some of these things are independent of one another, you can get things going faster. It's not that these streams are completely parallel—there are actually a lot of connections among these streams.

But you could recognize me in this room and you could recognize me outside, right? So recognizing me could be independent of where I am. If one process figures out who I'm talking to and another process figures out where that person is, you can do that in parallel and get it done faster. That's the idea of parallel processing: it helps with efficiency.

HB: What is the third stream that you mentioned before?

KGS: This is a hypothesis still. The main streams have been found as a result of neuropsychological studies and also from anatomy. The whole occipital lobe and parts of the temporal and parietal lobes are made of smaller components called areas, and each area actually has a representation of the visual world.

HB: How big is an "area", roughly?

KGS: In humans these are about a centimetre or two. The areas will be about half a centimetre on the surface of the brain. The biggest area is the primary visual cortex. It's about three centimetres big.

These areas are interconnected to each other. Physiologists and anatomists know which connections are ascending—do forward processing—and which connections are descending—do feedback processing—because of how they connect inside the layers of the brain.

HB: Can you describe a little more what you mean by ascending and what you mean by descending?

KGS: If you think about a stream, and the stream is composed of components called areas, what we have to figure out is how many areas you have in the stream, and then how information gets relayed from one area to another area.

If you think about a very sequential system, it will go, let's say, V1 to V2—so V1 would be first visual area; V2, second visual area—then V3, V4 and so on. That's called forward processing or ascending; it goes from a lower area to a higher area.

HB: So the information comes into the brain from the outside—I see something, say. Then it gets processed along the way by different areas. That's what you mean?

KGS: Yes. But it turns out there are also connections that are feedback connections or descending connections. So a higher area in the hierarchy, which would be farther away from the occipital lobe anatomically, will also sometimes send back a signal.

HB: And it will send back a signal saying what?

KGS: We are still trying to figure this out. There are also connections to the other lobes of the brain.

But here is an example of a signal that might be sent back. We're sitting here in this room, and perhaps in this context you don't care what room you are in. But in other contexts you might care which kind of room you are in: if you're in the kitchen or the office you might be doing different stuff. So the idea of this top-down effect is that you might want to process different aspects of the visual scene in the context of what you are doing. This is why you have these feedback connections, to modulate the processing in the context of your task.

HB: OK. I have one other question before I move to some specifics: would it be possible that a justification for why you have these streams over different areas of the brain, in terms of robustness, is that, from an evolutionary perspective, if everything was concentrated in one area and I get hit on the head in that one area—

KGS: So you are asking if there is redundancy.

It doesn't really help you. The answer is, "*No*". There is some redundancy in the visual system, but if you get a lesion on the parietal cortex—people who have a stroke for example—you might become unaware of parts of your visual field and that doesn't necessarily come back.

If you get a lesion in the temporal lobe—and we've studied patients like this—you might become prosopagnosic, or face blind, if it hits certain parts of the region. They can see the parts—they are not blind—but they cannot recognize faces.

HB: OK. Let me ask more specifically about this processing stream, the way it can be modelled, and more about your particular research.

What specific things have you done and are you working on to be able to model that process and these different processing streams?

KGS: Well, the first thing to do is actually measure them. I started doing fMRI in 1996 and at that time we were basically working off a monkey model of the brain and were able to measure V1, V2, V3, V3a, and half of V4. Rafi Malach had discovered a region that's involved in object processing: the lateral occipital cortex. People had also discovered, at that time, a region for motion processing: MT. People were just trying to map out the organization of the visual system.

HB: I'm hearing you say V1, V2, V3. I'm guessing that if you look very carefully at my brain you won't see a little 'V' and a '3', and a 'V' and a '2'. How do you even decide, first of all, what it is, and second, where it is?

KGS: First of all, we don't all agree. This is where science becomes interesting. You asked me before, *"What defines a visual area?"*. And there's a very famous paper in the field, written by David Van Essen that comprehensively formulates what we mean by "a visual area".

Essentially, there are four criteria for defining a visual area.

The first is the anatomical organization in terms of its cell make up: how they're organized across layers of cortex and how the particular sizes of cells and organizational structure vary across the corresponding cortical sheet. So that's an anatomical marker.

The second marker is the connections. If I'm V1, where do my axons send information, to which areas?

The third criterion is function. Maybe I'm an area that is specialized for doing a particular computation. For example, given certain visual stimuli, you might want to extract what colour it is separately from its identity, whether it's moving to your right or to your left and so on. There are regions that are specialized for extracting particular visual information—stereo depth information, or motion information, or colour information, or object shape information.

The fourth feature of a visual area is that it actually has a map of the visual field. It turns out that there are a lot of little maps in our brain.

HB: I have no idea what that means, so tell me more about that.

KGS: Well, your retina is kind of like a digital camera: each point in this visual space actually gets mapped to a photoreceptor in the retina. So basically there is a one-to-one relationship—because of the optics of the eye—between where something is in the world and where it hits the retina. And it turns out that there is a one-to-one mapping between certain photoreceptors in the retina and certain locations in the cortex. If I ask you to fixate on something, so I know exactly which part of the visual field gets projected on your retina, there is an exact replica of that in your brain; that's a map. It's a one-to-one map. It's a little distorted because the centre of your visual field is bigger than the periphery of the visual field, because you have more photoreceptors in the centre of your gaze than in the periphery.

So you might think, *Okay, I have a map of the world in V1. I'm done.* But this map gets replicated. So V2 also has a map of the world, and V3 has a map of the world, and V4 has a map of the world. This is what we can do very well with the functional imaging: have subjects stare at a dot as we project visual stimuli in different locations in the visual field and map it to the brain. It takes about 5 minutes to do that.

HB: What's going on when these maps are being replicated? Why do you think they're being replicated?

KGS: We're talking about processing hierarchy. It turns out that your photoreceptors are very small. Each photoreceptor sees just a tiny dot in the world. In visual science this is what's called a receptive field: what part of visual space gets activated. Each neuron codes a particular part of visual space. What happens is that across the hierarchy, in the early visual areas, each neuron really sees a tiny portion of visual space. You were talking about yourself moving in a scene; when that happens, a neuron in your V1 doesn't see the whole scene. It just sees a tiny portion of the scene, like you're looking through a keyhole.

As you ascend the hierarchy the receptive fields become bigger until at the higher end of the hierarchy they see whole objects, or larger parts of the scene. But to keep track of where information is coming from, having things mapped in a consistent way is really a very efficient way of organizing the information in a very systematic way.

HB: How is this linked to memory? Presumably if I see an object and I form an image of the object, at some point I'm going to be able to say, "*Oh, there's Kalanit. I've seen her before*", or "*This is a chair. I know what a chair looks like*". Clearly I'm doing something other than just being a sheet on which stimuli are being inputted. I'm somehow recollecting it to something else. How does *that* work? What's going on there?

KGS: There are multiple memory systems in the brain and some of them are outside the visual cortex. There are also varying degrees of memory. You could say, "*Oh, I've seen this person before. He looks familiar to me*". Or you could have an explicit memory—that you know who that person is: recognition. Or you could have a more elaborate memory like, *Oh, I met this person on that day and it was raining and stuff like that.*

HB: Wow, what a hard field you have. Just tying things to memory is hard enough, I should think.

KGS: Well, we don't deal with all those kinds of memory. We only deal with memories for things like recognizing a cat, for example. I need to have some knowledge about what a cat is.

HB: Philosophers have been arguing about what a cat is for a long time. But anyway...

KGS: Well, if I were to show a picture of a cat to a two-year-old, the two-year-old will probably be able to say, "*This is a cat*".

HB: If she's not a philosopher.

KGS: Yes. If she's not a philosopher.

This is the kind of stuff that interests me. How do you get a static picture of something and recognize what it is? Those kinds of memory are probably more visual in nature and don't really require all the semantic knowledge about what a cat likes to eat, and whether or not I saw a cat yesterday, and so on. I'm not going to deal with all these other memories.

But back to the processing streams. The idea of a processing stream is that, first of all, as you go up the series of processing, you process information over bigger extents of visual space.

And second of all, the processing becomes more elaborate. What do I mean by that? In the early stages, the neurons process local information about things like changes in contrast, or luminance, or little orientations. And as you go up the hierarchy the neurons process more elaborate information like the shapes of things.

HB: You don't process that in the beginning? So in V1 or V2 shape doesn't even kick in?

KGS: Well, some features do kick in, like orientation, or contrast, but these are all done locally.

You asked me before, "*Why do I need a processing stream?*" One of the ideas behind a processing stream is that from one stage to another there are different kinds of transformations until, at the end of the processing stream, you get something that would be useful for the goal of your computation.

V1 is the input to all kinds of visual processing: whether you want to extract motion information, or where something is, or what it is. It needs to collect content that would be useful for everything.

But as you go up the hierarchy, you become more specialized and you organize information that is particularly useful for the computation that you want to do. The information that you might need for recognizing something is different than the information you might need to figure out where something is moving to.

HB: And you can get some experimental evidence for this because when you do your experiments with people in the fMRI machine they're focusing on shape, they're focusing on movement, or they're focusing on different images, and then you see different areas of the brain being activated. Is that right?

KGS: Yes.

Questions for Discussion:

1. How is one's methodological approach to scientific issues influenced by one's academic background and training? Might Kalanit be said to take a more "computational systems" approach to vision than someone who might have come to the field from a direction less strongly aligned with computer science? Would it be fair to say that, at some level, she believes that the brain is a "very complex computer"?

2. In what ways is the notion of "redundancy" mentioned in this chapter related to "neuroplasticity"?

3. Kalanit's view of how the brain engages with the world through our eyes appears to be one of processing sensory inputs from the external world. How, if at all, can this be reconciled with other competing interpretations about the brain actively making predictions and then comparing those predictions with incoming sense data (see, for example, the Ideas Roadshow conversations **Minds and Machines** *with Miguel Nicolelis and* **Constructing Our World: The Brain's-Eye View** *with Lisa Feldman Barrett)?*

IV Experimental Evidence

Many discoveries; even more to do

KGS: Let me describe a very simple experiment. Suppose you read this study by the very eminent neuroscientist Bill Newsome, who studies neurons in the region called MT. Neurons in this region respond more when stimuli move than when they don't move—their firing rate increases. So what would I do if I wanted to find this region in a person's brain?

You find a volunteer subject and you create a very simple experiment. You take a bunch of dots and you show them, unmoving, for, say, ten seconds. Then you take the same bunch of dots and you start moving them around. You're a scientist, so you're going to repeat this experiment. Still again, and moving again. Still, moving. You do this six times. It takes you three and a half minutes.

Then you take your fMRI data to your computer and determine, of all the voxels you've recorded, which ones show the highest signals when the dots moved compared to when they didn't move. And you can determine which anatomical region is most strongly correlated with this phenomenon. It took you three and a half minutes and you've found the answer to that question.

Of course you can do a similar experiment to determine which part of the brain processes shapes. You take random shapes and then you blow them up into pieces and see which parts of the brain respond more to shapes than other things.

Or if you think there are areas in the brain that might be specialized for, say, recognizing faces, because they are evolutionarily important: you might show people pictures of faces compared to pictures of objects and see if any areas in the brain respond more strongly for faces than objects.

HB: And you would of course repeat these experiments, not only with that one particular volunteer, but you would do this over and over again with a wide variety of people.

How much variety is there in the brain anatomy between different people, and are there any additional correlations related to gender, age, DNA, and so forth?

KGS: It's unclear. There is a big Connectome project going on right now where they're scanning roughly a thousand people. One of our colleagues from Washington University in St Louis, David Van Essen, took twin brains and he thought that he would be able to recognize them by their anatomy, but it's not that obvious.

People thought that the anatomy could only predict the early sensory regions. Because these areas are connected to the peripheral nervous system, you know how it's wired, so it always has to be in the same anatomical location. People thought that all these higher-order areas were very variable between people.

In the research I've done, mainly with one of my students, Kevin Weiner, we found that the anatomy is much more predictable a function than you would think, even in these very high-order areas that are supposed to be moulded by experience and so on. So we've become better at actually predicting from anatomy where functions are.

For example, there is a particular gyrus—which is like a hill—but in the hill there is a particular little valley—it's called a sulcus, a mid-fusiform sulcus—and it's actually not in any of the anatomy textbooks. If you look at an anatomy textbook, it's not there. It's not documented. Kevin and I stared at 158 hemispheres and after seeing it over and over again in everybody's brain, we understood that it really exists.

HB: Have you told these anatomy guys? Are the new textbooks at least going to show it?

KGS: Are they going to show it? I don't know.

Kevin is interested in the history of science. In Stanford there is a library that has anatomy books from the 1800s in its archives. He was really bothered saying, *"Why is it not in the books?"* He figured maybe they didn't see it. What seemed to have happened is that, because it's a small sulcus and they had just one example, they missed it.

The big sulci are going to be present in everybody's brain, but I don't know how much the small sulci are idiosyncratic of a particular brain, or not. One of the advantages of doing neuroimaging is it's not invasive, so you can scan a lot of people and you can get a lot of data. If you start seeing something over and over again in every single subject and you can start quantifying what's variable and what's not, then you can start making sense out of it.

HB: According to what you're saying, there's a significant amount of variation in terms of the particularities of the sulci and the folding and so forth, but then there are some—typically, but not exclusively, the big ones—that everybody has.

So my sense is that there is a lot of local variation but then there are some things that are very constant—getting back to your example earlier about most people having five fingers—and there has been a lack of attention paid—or a lack of awareness perhaps—to some of the smaller ones that are actually present in all humans. Is that a fair summary?

KGS: Yes. Basically the general organization is going to be present in everybody's brain. For example, everybody will have these regions, the main sulci, and stuff like that. There are particular characteristics that vary across subjects. Sometimes a sulcus will have two branches or sometimes it will have one branch, sometimes it will be very long and very deep, and sometimes it will be shorter. When you look at the brain you have to train yourself to recognize the same features despite this variability.

What you said was very accurate though. The kind of work that's happening in my lab as well as other labs is that we're trying to characterize this variability.

And then you can start seeing the regularities; and once you start seeing the regularities you can start making new inferences that you haven't been able to see before.

HB: This region that you and Kevin found, this thing whose name I've forgotten—

KGS: It's a "mid-fusiform sulcus"—it's like a hidden valley that we found in the middle of a well-recognized area called the fusiform gyrus.

HB:—this mid-fusiform sulcus, then—this hidden valley—that nobody seemed to notice until now but is apparently universal—I'm guessing that the thinking is that, since everyone has one of these things, there is some central common processing going on there, right?

KGS: Yes. We found two things that make it interesting.

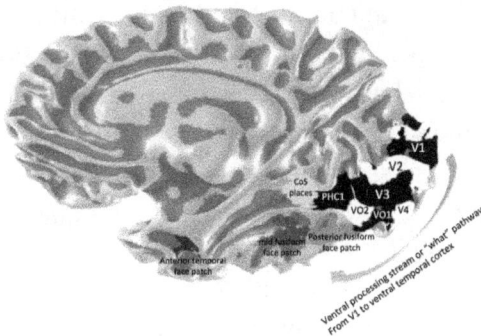

You see these red splotches here on this brain? All these red patches are regions of the brain that respond more strongly when you see people's faces compared to other visual stimuli. It turns out that everybody has them. So there is some dedicated hardware in your brain that seems to be involved in processing faces.

This discovery was made by Nancy Kanwisher in 1997. Initially she thought there was just one of them. She said there was a module in the brain—she described it as a blueberry-sized module—that only processes faces. That was 1997. So she was the first person to discover this with fMRI.

What we've found since then, with our methods getting better, is that there are a number of these modules that are organized in a very systematic way. And what we've recently found is that they are actually very predictable by anatomy. They happen in every person's brain in the same part of their brain. If you find this little sulcus, you always find this face-selective region on this sulcus.

We started another collaboration with people who look at the histology, the anatomy makeup of the brain. You cannot do this in a living subject because you have to do it on slices of the brain under the microscope, so they use post-mortem brains. And they discovered that different regions have different hardware.

You were suggesting a moment ago that, if there is specialization for face processing, maybe it will require very dedicated hardware that's optimized for doing this kind of computation.

We cannot map function to their data because they were using post-mortem brains. But it turns out that the location where they find this specialized hardware is very aligned with this mid-fusiform sulcus that we found. So this gets us thinking about how some type of specially-dedicated hardware might lead to some specific processing that's relevant to our perception. We are still finding our way there. But that's our plan.

HB: As you were talking I thought about some Oliver Sacks-type of scenarios. There must be some people out there who can't recognize faces, or have problems remembering faces. I can imagine that you might have subjects who have, for whatever reason, difficulties with facial recognition whom you can then put in one of these scanners and maybe see some anomalies that don't match up to those regions that you were just talking about.

KGS: These people are called prosopagnosics and their condition is sometimes called "face blindness". Sometimes it's because they're missing a part of their brain due to some brain injury, and sometimes they don't have any atypicalities in their brain but they're still very poor at recognizing faces, and we've just started looking at that.

It looks like they *do* have these regions but they're smaller. But I haven't looked at enough people to really tell you what's different about their brains. It's not very obvious.

HB: How many of them are there?

KGS: We're not sure how common it is, but some places are relatively well-documented. There's a study from a group in Germany that examined a village of 700 people and estimated that 2% of the population is really bad.

HB: 2%!? That seems pretty high to me. What's going on in this village?

KGS: It is high. So these people exist.

HB: They certainly exist in that village, it seems.

KGS: They exist in the world. Most of the time, though, I'm focused on trying to figure out what happens in typical people.

Questions for Discussion:

1. Are you surprised at the notion that some small areas of the brain might not be mapped out in all the anatomy textbooks? What does this imply about both the anatomical complexity and variability of the human brain?

2. How do you think it is possible to assess "different types of hardware" in post-mortem brains? What is meant by "hardware" under these circumstances?

V. A Startling Result

Stumbling upon specialized hardware

KGS: As I was saying, most of the time we're taking non-invasive measurements of typical people. However, once in a while we have the opportunity to record directly from the surface of the brain. This sometimes happens when subjects get evaluated for surgery for epilepsy.

There is an epilepsy clinic here at Stanford and the neurologist is Josef Parvizi. These patients typically have intractable epilepsy that can't be treated with medication, so the doctors try to evaluate where the seizure starts and whether or not it might be possible to try to reset that little part of the brain to help them with their epilepsy.

They come to Stanford for about a week and they have electrodes implanted on the surface of their brain. For a week they are basically just sitting there waiting for a seizure to happen so the doctors can track where the seizure is. Sometimes, if they allow us to do so, we get to actually record directly from areas of their brain.

And as it happens, for one of these subjects, we did an fMRI before he went in for surgery, so we were able to map these face-selective regions in his brain.

Part of the medical procedure for epilepsy is to put electrical current through these electrodes to see if it disrupts brain activity because you obviously don't want the surgeon to remove some vital piece of cortex. So if that vital piece of cortex is going to make you unable to see faces, they're not going to do surgery.

What I'll show you is an experiment we ran with a subject here, and we were able to map these two face-selective regions on the fusiform. You'll see what happens when he's looking at the doctor's face; in some trials he receives an electrical current—you'll see how much current it is—and in some trials the doctor does everything the same but there is no current—it's a sham trial—so the subject won't guess.

HB: It's a control measure.

KGS: Exactly.

This patient's name is Ron Blackwell. When the surgeon implants the electrode grids, he doesn't see the surface of the cortex and just slips these electrodes under without actually seeing—he didn't have access to our MRIs. He had about maybe a dozen electrodes around the ventral surface and we were just really lucky that two of these electrodes were on the centre of these two face-selective regions, because these electrodes are spaced about a centimetre apart.

HB: Oh, really? So it was a complete fluke.

KGS: It was just a complete fluke. Yes.

You could say, "*Well, we know that in typical people they show higher responses to faces and other stimuli, but are they causally involved in processing faces*"? fMRI is all correlational. But this gives us a unique window into their causal involvement. If you disrupt their normal activity by applying a current, the question is, "*Will it create a specific deficit or will it create a general deficit?*" Maybe it will affect their face perception and other stuff.

You could also ask, "*What if we had electrodes over here; if you apply an electrical current there, will it affect the subject's face perception?*"

What I want to show you here is what happens when Ron gets stimulated by these electrodes. Dr. Parvizi is asking him to look at his face as he's doing all this. It begins and ends with a sham, which is no electrical current, and then there are two trials where a current exists.

Start of Video

Electrical stimulation of fusiform face-selective regions in a patient implanted with intracranial electrodes

Dr. Parvizi: Just look at my face and tell me what happens when I do this. Alright?

[Sham]

Ron: Nothing.

Dr. Parvizi: Nothing? Okay. I'm going to do it one more time. Look at my face. 1, 2, 3.

[4 mAmp]

Ron: You just turned into somebody else. Your face metamorphosed. Your nose got saggy, went to the left. You almost looked like somebody I'd seen before, but somebody different. That was a trip.

Dr. Parvizi: Could you see my eyes in place?

Ron: I could see your eyes, but you could have been somebody else who—well, he had similar eyes to you, but you were someone else. Your whole face just sort of metamorphosed.

Dr. Parvizi: Metamorphosed. Did it get warped?

Ron: Yes.

Dr. Parvizi: Did it keep its own features, or shape?

Ron: No. It's almost like the shape of your face—your features drooped. Sort of like they drooped to a particular direction. Kind of stretched and drooped.

Dr. Parvizi: All right. Shall we try again?

Ron: Okay.

Dr. Parvizi: 1, 2, 3.

[3 mAmp]

Ron: Yeah, it metamorphosed again and you look like someone I've seen before. But maybe a different person in my memory. Almost like your nose kind of shifted to the left a bit, and your look just changed.

Dr. Parvizi: Tell me more.

Ron: Well, I don't have a photographic memory, but you turned into someone else. You looked like someone else.

Dr. Parvizi: Did I keep my gender?

Ron: Yeah. Oh yeah.

Dr. Parvizi: How did you know that I didn't become female?

Ron: Because you were still wearing a suit and tie.

Dr. Parvizi: Oh, you could see the suit and tie?

Ron: Yeah. Only your face changed, everything else was the same.

Dr. Parvizi: Did the skin colour remain the same?

Ron: Yes.

Dr. Parvizi: Did the position of my lips, and nose, and eyes stay the same when they got warped?

Ron: They shifted. Let's say they shifted to a side and maybe stretched. But they didn't get larger or smaller. It was more of a perception. How I perceived your face.

Dr. Parvizi: Interesting. Tell me more.

Ron: That's about all I can say. All of a sudden you were you, then you weren't you. You could have been someone else who was standing there in front of me.

Dr. Parvizi: Shall we do it again? Ready?

Ron: Okay.

Dr. Parvizi: Tell me if the same thing happens again. 1, 2, 3.

[Sham]

Ron: Nothing happened that time.

End of Video

HB: Wow. So let's back up for a moment and talk about what this means. There are some regions that were hypothesized as being critical for facial recognition. They were stimulated and there was very suggestive evidence that, having stimulated these areas, there was a very strong causal link to our facial recognition perception. What sort of things might we be able to conclude from that?

KGS: The most important thing that we want to know is if there is some dedicated hardware that seems to be involved for a particular kind of processing. This shows very strong evidence towards the existence of such hardware.

As I told you, with fMRI I scan a person many times and I can show what kind of stimuli will increase activation in certain areas.

For example, we've run experiments when we show pictures in the flash of a second. Sometimes subjects can see it and sometimes they can't. We can show that these regions respond more strongly when there is a picture and the subject sees it than when there is a picture and you don't see it.

So we've shown in a correlative way that the activation in the region predicts what the subject perceives. But we were never able to show directly that it's necessary or sufficient. These kinds of experiments are very critical because they show that if you perturb these areas, it really affects your perception.

HB: And moreover you also show that, even when these areas *are* perturbed, many other things—like the doctor's suit or skin colour—remain unchanged.

KGS: There are two additional things that we did. In his room, Ron had a bunch of objects. Let's say a get-well balloon which requires reading, fine visual acuity. When we stimulated these electrodes, he was still able to read. So it's not that anything you stimulate will cause a visual distortion.

We also stimulated some more medial regions that seem to selectively respond to places and scenes more than to objects and faces. But when we stimulated these regions, Ron didn't get any visual distortion, neither of faces nor of anything else. So we have enough data to be very clear that there is very specific brain tissue that is involved in a very specific function.

And that's what's so cool about it.

HB: That whole thing about how he could see the suit and the tie, that was just remarkable, that it was so incredibly specific: it was just the face. He knew the gender. He knew the suit. That was really remarkable.

KGS: This is great evidence for specialization in the visual system. There might be dedicated hardware for particular processing.

And you might ask yourself, "*Why*?" The answer is probably because it has been evolutionarily important to recognize faces—as a species, we're a very social group.

Also, there are some aspects of a face that you probably wouldn't care about with chairs. For example, gender is something you don't care about with a chair. You may care about the age of a face to determine things like social stature and what not. There are a lot of reasons evolutionarily why we might have this dedicated software and why it would be important to do that sort of processing very quickly.

Questions for Discussion:

1. What, exactly, do you think Kalanit means when she talks about how "fMRI is all correlational" but studies like the one in the video uniquely allow a determination of causality?

2. What role did the sham trials play in establishing necessity or sufficiency regarding the current theory of a specific brain region for facial recognition?

3. What sort of experiments would you perform on prosopagnosics if you were to find yourself in possession of an entirely safe and non-invasive machine that could perform the same sorts of experiments as the one in this video?

VI. Neuroplasticity
Assessing flexibility

HB: Let me tack a bit in another direction now and ask you about the plasticity of the brain. We've talked quite a bit about localized function, specialized hardware and processing streams.

If I'm some guy going through *Scientific American* or something, I may have heard all this talk about how the brain can change and adapt, how it is, to some extent anyway, plastic: there's some localization clearly, but it's not as if neurons are always doing exactly the same sorts of things all the time. How does the notion of plasticity fit in with your particular work? Are you consciously thinking about that as you move forwards with developing these particular models?

KGS: We're actually thinking about plasticity a lot in multiple domains. One domain is actually in the domain of face recognition. You constantly learn new faces throughout your lifespan, so there has to be a way to constantly update your day-to-day experience and updating your representation of faces.

We know that these can modulate brain responses. For example, if I show a subject the same picture over and over again I get a decreased brain response because the first picture was new and then by repeating the same picture you get a decreased rate of response because the brain is updating. That's a kind of plasticity, but it's short-term plasticity. It doesn't really involve changing these regions.

But they do change overtime—another kind of plasticity that we're looking at in my lab specifically is plasticity over the lifespan. In particular, we're interested in how high-level visual areas develop from childhood.

We've studied seven-year-olds all the way to adulthood and what we've found is that, in children, the general topology, the layout of these regions, is very similar to adults. It's like they have the same kind of terrain but it gets more chiselled with experience.

Basically what happens with kids is that they have the same regions but they're smaller and they're less selective. One of the things we've found is that in adults you get differentiated responses for adults' faces and kids' faces, but in kids it's like they have one category for faces—it's not that differentiated between kids' faces and adults' faces.

Once again, experience can mould your representations and, as you become more expert, and maybe you become more interested in your own group, you get more distinct representations.

The third kind of plasticity that we're measuring is what might happen because of, say, brain damage. Suppose you had a stroke in the occipital area—will it affect downstream areas? Will they get reorganized because now they are lacking input from an earlier area? These are the three kinds of plasticity that we're looking into.

HB: Have you had a chance to theorize or find any data with regard to the third kind, in terms of restructuring?

KGS: We're starting to look into that.

HB: With the second kind, I'm trying to imagine someone who has an overwhelming necessity of being able to rely upon facial recognition, maybe someone who, in their job, is doing that sort of thing constantly. I'm trying to imagine if such people exist and, if so, could you study them?

KGS: I think that would be really interesting.

Two scientists—Brad Duchaine at Dartmouth and Ken Nakayama at Harvard— have designed this test called the Cambridge Face Memory Test (CFMT). You can do it online. They'll show you, for example, six faces and then they'll show you these faces among other faces and you have to say which of the faces you've seen before, and they'll show them in different illuminations and with different amounts of noise.

They've developed this battery of tests to predict how good people are at face recognition. You asked me, *"How do you know that 2% of the population is bad?"* About 30,000 people have taken this test online, and people have done it across continents. We have good numbers now, so we can identify a normative range. It turns out that, just as there are these 2% of people who are really bad, there are people at the top of the spectrum who are super-recognizers.

We've never done this—I've talked a little bit with Brad about prosopagnosics—but it would be really interesting to see what's actually different about people who are really good.

HB: They don't live in the same town in Germany by any chance, do they?

KGS: I don't know. I haven't been to that town. Nor can I pronounce the name of that town, so don't ask me.

Anyway, I think that would be really interesting.

In my lab, we're just launching a very long longitudinal study in which we'll study a wide range of different ages—5–7-year-olds; 7–11-year-olds and 22–23-year-olds and follow them across 5 years. Then we'll be able to measure how their brain changes over time.

What we hypothesize is that corresponding changes in the children will be bigger than the changes in the adults. There are some points in time when you're going to learn a lot more faces, and they become a lot more socially relevant to you.

But, of course, we don't know. This is why we're interested in doing this longitudinal study, because so far we've only done it cross-sectional and this will reduce the between-subject variability by following the same subject over time.

HB: Sounds reasonable. One more question before I leave plasticity. Since you're studying vision, what about people who are blind, who are either born blind or become blind? Have you had the opportunity to do any studies, or do you know of any studies whereby you can see evidence for plasticity in the vision processing centres of the brain?

KGS: This is a very good question. I don't do any studies on blind people. There are a bunch of groups throughout the world who study the congenitally blind.

I told you before that 30% of the brain is used for visual processing, so what happens if you have no visual inputs? Does it die? Is it not used?

It turns out that there is a lot of recycling. It turns out that, in the blind people, the visual cortex is used for other things. The earliest study was done by Alvaro Pascual-Leone and colleagues. They showed that, the same region I showed you, V1, gets activated in the blind people when they read Braille, compared to when they just sweep their hand over something that doesn't have Braille.

HB: Really? So the tactile sense somehow gets linked to vision.

KGS: People have done more experiments and it looks like it's related to the top-down stuff I was talking about earlier. It's not that the different sensory areas get reconnected. Stuff that's involved in verbal memory and language gets recycled because you don't have the sensory input, but you still have the top-down stuff.

Alfonso Caramazza and Brad Mahon have actually made claims that there's specialization—like I showed you for faces and places in sighted people—in blind people. So it looks like some parts of cortex might be more plastic than the higher-order levels. These groups are looking into that. The key question is, *How much is pre-wired and how much does your experience mould these regions?*

Questions for Discussion:

1. Which specific jobs or activities do you think might naturally train your facial recognition skills more than others?

2. What would you hypothesize is happening with those who have poor facial recognition skills? Do you think that they are missing some of the required "hardware" or would you guess that the problem lies more with their ability to make sufficiently strong connections to process the information? Might it be possible, under the right training regimen, for those people to improve their facial recognition skills given the brain's inherent plasticity?

VII. The Road Ahead

Better measurements, better models, deeper understanding

HB: This has got to be one of the coolest fields in science right now. There is so much going on, not only in terms of ideas and interesting people involved, but in terms of data, correlations between theory and experiment, and so forth.

My sense is that we're really in a golden age of cognitive science. Is this the sort of thing that you had imagined would be the case when you started moving into the field? Is it moving faster than you had expected? Are things as sunny and exciting and dynamic as I, an outside observer, am saying they are?

KGS: I was very fortunate to be in the right place at the right time. As I told you earlier, fMRI just kicked in when I was doing my PhD. It was this whole new thing. When I heard Rafi come and talk about it, I realized immediately that this was going to be a very powerful technique. It has clearly revolutionized cognitive neuroscience.

If you were to go to a Society for Neuroscience conference in 1997 or 1998, there would be roughly eleven sessions focused on vision—one session about humans and ten sessions about animals. Today these numbers are flipped because a lot of people have realized how powerful MRI is.

If you're the first one jumping on the wagon, it's easier to make discoveries than if you're the hundredth person jumping on the wagon. For me, the thing that neuroimaging changed the most, is that we have a clearer sense of what we can accomplish, and I've become even more intrigued by how complex the visual system is, and how much more there still is to understand.

When I was in grad school, in the model that I was taught, the hierarchy had four stags: V1, V2, V4—I'll tell you why V4 and not V3 in a moment—and IT: inferior temporal cortex. So if I just think of four black boxes, I'm done. Now, sixteen years later, this IT black box turns out to have at least ten subregions.

So on one hand, I'm more modest because I know the model from fifteen years ago is not valid, but now we have better tools to actually start tracking and making models of these subregions. I think we will continue to advance the field more substantially now just because we have more knowledge than fifteen years ago.

HB: Why no V3?

KGS: This is a speculation. In humans, as I showed you, V2 and V3 are about the same size, while V1 is a little bigger. In macaque monkeys V3 is very small and second of all there are connections directly from V2 to V4 that bypass V3. So because it's small and because it's unclear that having another level is actually going to change the model much, people just stick with V1, V2, and V4 and then IT. But I think that's wrong. I think we need a V3 there, in humans for sure.

HB: You can always relabel V4 V3.

KGS: It's in the model. V1, V2, V4.

HB: Well, it's your model.

KGS: It's not my model.

HB: I mean, it's the field's model. It's your professional model.

KGS: It's my professional model, yes.

HB: Not only has the field changed in terms of a rapid accumulation of scientific results and a corresponding increase in our level of understanding, but I'm guessing it's also been transformed sociologically. You're a faculty member at Stanford, a vision specialist with a background in computational neuroscience, and you're also a faculty member in the Department of Psychology.

Back in the days when I was an undergraduate, I think it's safe to say that no faculty members in the Department of Psychology would be like you. It seems to me there has been this explosion in the field of cognitive science that has integrated people from across a wide variety of fields and activities. Has it been easy, do you think, to integrate all these people? What does it feel like for you being a professional psychologist? Do these distinctions make any sense whatsoever anymore or have things just changed entirely?

KGS: In our department we have a specific area that's called neuro-science. So there is neuroscience and psychology. The neuroscience in our department existed even in the 60s, I think. So neuroscience has not really been completely novel to psychology. It's been there for a while. A lot of psychology departments did have animal studies, and these studies used neuroscience. Sometimes neuroscience is included in med school and sometimes in psychology departments.

But I do think that around the 60s and 70s there was a big revo-lution in psychology in which cognitive psychology became the core, where all the action was. Today a lot of the major discoveries are happening in neuroscience.

But neuroscience has infiltrated all fields of psychology. You can use neuroscience to study personality, you can study depression, you can study social perception. There is a field called social neurosci-ence. It affects things that are at the core of cognitive science, like decision-making. I don't know if that answered your question or not.

HB: Well, it wasn't a very well-posed question, so let me try to ask another one, which will hopefully be a bit more focused.

Are there some psychologists who have their nose out of joint and say, "*All these bloody fMRI people—every single thing is now fMRI. I'm studying why people are tremendously upset at life because they couldn't live up to their father's expectations and now they're making me go into an fMRI lab. To hell with them! Why don't they get their own bloody department and stay away from mine!?*"

That's going a bit over the top, I appreciate, but I was trying to be more focused. Anyway, in your experience is there a sense of resentment amongst people who do what might be considered more "traditional psychology" because now all of a sudden everything is cognitive neuroscience, at least they are feeling that sort of pressure?

KGS: The field of psychology is definitely not all cognitive neuroscience. The field of psychology is a very well-grounded field. It studies the effects of psychology, personality, social interactions. These fields are still very central to psychology.

I think when you have people in a group there is always somebody who's going to feel some resentment for multiple reasons, but I don't really feel that's been an issue for me or in the field in general.

HB: OK, let me move now to the future. I'm guessing you're very optimistic about the future. I'd like you to comment on your level of optimism as well as the various research avenues that you would like to pursue, and what some of your colleagues are doing that you are excited to learn about.

KGS: I'm a very optimistic person. I don't think you can continue to do research if you are a pessimist. I've been very fortunate that I've been working with really terrific people. Without all these different people I wouldn't be making these discoveries. Our work is very much collaborative. If you are doing some theoretical work you can sit in your office and theorize, but experimental work really needs a good group of people who are very motivated and dedicated and I've been very fortunate in that regard. And I think we're going to see even more talent coming into the field in the future.

More specifically, there are several directions that are going to be very important for me personally. First, these technologies continue to evolve. We are still trying to improve the temporal resolution of these measurements, the spatial resolution of these measurements, as well as integrating between the different types of data—I mentioned functional MRI, anatomical MRI, trying to figure out the wiring of the brain—and we are actually trying to figure out how we can get to the circuit level: looking at the cell makeup of these regions. That would be very cool. Some of it is science fiction.

HB: For now.

KGS: Right—some of it will happen, because there is a whole field of people who are working on improving the measurements, the technology.

The second kind of thing that I think is happening—which is why it's a really exciting time to be in vision science—is that people have really improved the modelling.

Since we have a lot more data, we can start generating a new kind of model—I call them generative models. The idea is to model the neurons in the different stages of the processing stream, figuring out the computations that they do and what kind of information they relay to the next stage—which is a more traditional model—as well as having them predict how my brain responses should look for a new kind of stimuli and actually validate that with fMRI.

We've started dabbling with this, and I think this will be another major breakthrough in the field: when we have taken this knowledge, actually predicted specific brain data, and then directly measure our prediction. I think that would be extremely cool.

HB: Do you have any sort of albeit speculative sense of what these models might look like?

KGS: At their core these models are about what information is transformed from one stage to another, trying to figure out what kind of optimization principle the brain uses. This is what we're trying to figure out. *Why is the brain so organized? Why is it so similar across my brain and yours?*

We think that there is some kind of basic computational principle that the brain employs to optimize how efficient things are, how robust things are, and also which kind of information you want to integrate and which kind of information you want to segregate. If we get a handle on exactly what kinds of information are important to lump together and what needs to be kept separable, then I think we're going to make a big breakthrough with these kinds of models.

HB: That's great. Anything I missed? Anything you'd like to add?

KGS: I can go on like this for days, or weeks.

HB: I'm being quite serious—feel free to add whatever you'd like. The Stanford police have probably already dragged my car away by now anyway—I think I parked illegally.

KGS: Well, there is one thing I forgot to mention earlier when you asked me about the super-recognizers. We've also collaborated with a psychiatrist here at Stanford, Allan Reiss. He studies a variety of people who have genetic disorders, and something called Williams syndrome is one of these disorders.

It's a deletion on chromosome seven, so it's kind of like Down syndrome in the sense that these people are born like that. It's also like Down syndrome in the sense that those with Williams syndrome also have a characteristic face, kind of an elfin face.

The reason we became interested in this disorder is that people with Williams syndrome are not very good with spatial vision, but they're really awesome at remembering people's faces.

So they'll see a person once—in a supermarket, say—and five years later they'll say, *"Oh, I've seen you!"*. Their cognitive abilities have some strengths and some downsides. They don't have very good spatial vision, or spatial comprehension, but their language and music abilities are very good and their facial recognition is great.

We were interested to see if this is something that is relatively good despite other things being different about their brains, or is it that they have something different about their brain that makes them even better than typical subjects?

We've done fMRI experiments with adults who have Williams syndrome. We put them in the scanner and show them pictures of faces, of objects, of places, and some random patterns. You find the same regions in the fusiform that show higher responses to faces compared to non-faces. I did this with a postdoc in my lab, Golijeh Golarai.

What was interesting about those with Williams syndrome is that their brain is a little bit smaller—about 10% smaller on average than the typical person. And this fusiform gyrus that I described to you before—this is where you would normally find these face select-ive regions—is also smaller in those with Williams syndrome. But if you look at their brains, they have more parts of their temporal lobe involved in face processing than is typical, and more so if you accommodate for the fact that their fusiform is smaller.

Maybe a quarter of my fusiform gyrus is devoted to face process-ing, but in the people with Williams syndrome it's much more. A bigger expanse is devoted to processing faces. Are they born like that or is it something about their experience?

They're interested in faces throughout their lifespan and they basically just become more expert at it. The way my colleague Allan Reiss thinks about it is that when you're a baby you're naturally attracted to faces but at some point you start getting interested in other stuff like how to manipulate objects, or other kinds of baby stuff.

But maybe what happens with those with Williams syndrome is that this interest in faces doesn't get shut down in the same typical way—they *don't* lose interest in faces and start getting interested in other things. And because of this, their experience is more driven by faces than a typical subject and therefore they develop more brain hardware that's devoted to coding faces and that gives them their extra ability. So it's not necessarily the case that their brain has been wired—

HB: No, it's the nurture argument.

KGS: Yes. But this is an example of how nature and nurture might interact to shape neural representation in the brain. So it happens in the same place, but how much of it, or how specialized it becomes might be dependent on your lifelong experience.

HB: And if this hypothesis is correct, then it seems to me that's clearly another argument for plasticity, because their brain is changing in a way that other brains at that age would not be. That is to say, they're emphasizing this action over another and that's having some neurophysiological consequences and so forth. Well, that was great. Thank you very much Kalanit.

KGS: Thank you.

Questions for Discussion:

1. If we were to develop a comprehensive understanding of the brain, would there still be a need for "traditional psychology"?

2. Kalanit mentioned that Williams syndrome was linked to a specific deletion on chromosome 7—what role do you think genetics will play in the development of a deeper understanding of neuroscientific mechanisms? (Readers interested in this topic are referred to two Ideas Roadshow conversations: **Our Human Variability** *with University of Toronto geneticist Stephen Scherer and* **Autism: A Genetic Perspective** *with UC Irvine geneticist Jay Gargus.)*

3. Are there aspects associated with the rapid development of neuroscience that trouble you? Should bioethicists be involved in establishing more regulations and guidelines? If so, how and in what way?

Investigating Intelligence

A conversation with John Duncan

Introduction

Thinking Deeper

What if you could take a pill that would make you more intelligent?

Science fiction? Well, maybe not. Researchers have been studying intelligence for some time now and have begun to make significant progress. But it all starts, of course, with understanding what, exactly, we mean by intelligence in the first place.

In his popular book, *How Intelligence Happens*, University of Cambridge cognitive scientist John Duncan sets the stage by re-introducing us to the work of the influential English scientist Charles Spearman.

Back in 1904, Spearman discovered that a clear correlation existed between people's abilities to perform various tasks. Those who were good at one sort of thing tended, in a mathematically quantifiable way, to be good at others. The wider the sample of tasks, the clearer this correlation became. This led Spearman to develop his theory of a so-called "g factor" or "general factor", reflecting a certain aptitude that some people had which enabled them to be successful across a wide range of activities.

Intriguingly, some tasks proved to be fairly reliable indicators of this factor almost all by themselves: Spearman applied his statistical approach to the work of Alfred Binet and others, and the modern "IQ test" was born.

But what does it all mean?

Well, it's hardly straightforward: the road to a more rigorous understanding of human intelligence is fraught with difficulties.

The first is sociological. Many are concerned that such investigations are driven by a dangerous sense of elitist eugenics that seeks to extend itself far beyond any strictly defined sense of cognitive ability.

A common criticism levelled against IQ tests, for example, is that they are motivated by an insidious desire to reduce all human experiences, and all humans, to a single number.

John unhesitatingly dismisses this idea straight away:

> *"Right from the very beginning of the scientific study of cognitive abilities it's been perfectly clear that you cannot assess somebody's abilities, and certainly not their worth, through a single number. In fact, there are an infinite number of things that are important to people and we value them for so many different characteristics: their honesty, their good humour, their ability at math, their ability to play football and so on—an infinite number of different things. That is the truth; and if you think you can explain everything important about somebody with one test, it's obviously doomed. As far as I know, nobody has ever thought that, although many people are criticized for thinking it.*

> *"However, what started with the work of the British psychologist Charles Spearman in the early part of the 20th century was the study of something much more specific than that. That was an empirical discovery drawn from experiments on measuring people's ability to do things; and putting forward a theory to explain it."*

The problem is, of course, that the scientific study of intelligence has a nasty habit of running smack into our pre-conceived egalitarian principles.

> *"It's very common to hear that everybody has their strengths, and that's a great thing. But then, meanwhile, you go into the job interview next door and people say, "I really like **that** candidate, **that's** the smart one." Well, what do they think they **mean** by that? They don't believe this person will do well at the job because of being good at this or that particular thing, but rather because they just somehow felt that they got more out of them intellectually.*

"I think people always have an intuition—which to some extent, at least, is true—that the same people tend to be able to flexibly address themselves to different sorts of problems. But what's important about a scientific program is that for a given question, such as predicting how well a person will do in a new job when you're interviewing her, this is something that you can actually measure."

But it's one thing to measure something and another to truly understand it. If this g factor is a real thing that somehow enables some of us to more flexibly address different sorts of problems, what is it, exactly? And where can it be found in the brain?

Through years of rigorous testing and close examination of brain-imaging technology, Duncan believes he's narrowing in on the answer. The key to g, he believes, lies in a very particular brain network that lies in the frontal and parietal lobes.

"Very interestingly, if you gave many different sorts of tasks—tests of memory or language, or even identifying faces or holding something in short term memory—this same network tends to be a part of the brain's response.

"So whatever g is, it's something that's important in organizing many different sorts of activity, which is just what you should think if it's going to explain the g factor. And much of what we're doing now is trying to understand what's going on inside these regions of the brain as problems are solved."

The key word here is "organizing". John is convinced that the primary role this network plays is to allow us to focus sufficiently on a task at hand in order to select the key thing that enables us to find the solution. That is, in a nutshell, what separates those with high g from those without it, and effectively what we mean when we call someone "intelligent".

"Focus is really the same thing as selection, if you go into it. It means that you take part of the problem and not the rest. Interestingly, I think this is also very closely related to what many people think of

as the heart of what's special about the human mind: the power of abstraction, being able to think abstractly.

"Often we tend to think of abstraction in, if you like, rather abstract terms: we think that abstract thinkers are philosophers or those who can tackle complex and recondite problems. I think there's probably a simpler way to think about abstraction as the ability to see the common important thread between a great many instances that are different from one another. This is what we mean by abstraction.

An instance of justice occurs, for example, no matter whether it happened in the court or on the playground. You can see we're now very close to the idea of attention or focus: we're picking up just one critical aspect of a situation and throwing away all the other things that differ between them.

"I think that this brain network we associate with g is very much related to this heart of human thinking: the ability to abstract out just which aspect of a situation is critical for the current moment."

Now that we have a clearer sense of what we mean by intelligence, together with where the networks associated with it can be found in the brain, might we finally find a way to increase it? Can we improve our own g factor?

"I'm certain that it's possible and I'm equally certain that we have very little idea at present how to do it.

"To me the most interesting thing is that no matter what experiment you look at, there's always a strong environmental component to it. That means that somewhere in a person's life, stuff did happen that really affected their scores on tests like this.

"Since the scores are so predictive of how well you solve problems in your everyday life, that implies that some kinds of experience do enable a person to learn better to solve the problems that they care about.

"And that is very big news, if we could only find a way to do it.

Far from steering us towards eugenics, then, John's research is pointing us firmly in the direction of better educational techniques.

That smart pill, in other words, might just be a better teacher.

The Conversation

I. Searching For A Definition

A curious correlation

HB : Since you're a scientist who studies intelligence, perhaps the most obvious thing to do at the beginning is to discuss what is meant, exactly, by intelligence and how we might go about rigorously evaluating it.

JD: Well, of course, people mean many different things by "intelligence", and there are correspondingly many different strands of scientific inquiry into its nature. One thing that is popularly known about intelligence—and is subject to much controversy—is intelligence testing: the attempt to use psychological measurements to assess a person's "brainpower", if you like.

There's the study of artificial intelligence, which has been active since the 1950s, a fascinating undertaking to develop thinking computers and examine the differences between the way they think and the way that we think.

There is a whole biological side to the study, where we look at brain mechanisms either through the effects of damage to particular parts of the brain and how it affects people's reasoning or cognitive powers, together with modern methods of brain imaging that peer inside the skull and see what's happening in the brain.

I think what's interesting and exciting about this moment in time is that we're beginning to see how all these different strands can be brought together and put into an overall framework of how it is that the mind engages in and solves the many complicated problems that we deal with everyday.

Every animal has its niche; and our niche is, essentially, our intelligence. What's made homo sapiens what they are on this planet is this ability to go into new situations, understand how they're structured, and then bend the world to work out in our favour.

I find this line of research absolutely fascinating. This is, I think, both the strength and also some of the risk of this field of study. We sometimes overvalue words like "intelligence" and "stupid" so that we make so many value judgments on the basis of simply those words. For example, you might unthinkingly say something like, "I hate this stupid weather, it's really annoying me", and I think that those sorts of value elements really get in the way of trying to investigate it as a scientific enterprise.

HB: So much so, in fact, that one could argue that, to some extent at least in the popular consciousness, the word "intelligence" has almost lost all meaning.

From a scientific perspective, one naturally wants to be rigorous, one wants to define what it is that one is talking about, but in terms of the man on the street, as you say, it's extremely unclear what one means by the word "intelligence". One could be talking about knowledge, one could be talking about intuition, one could be talking about insight. I think we all have some basic sense of what it is, but to study something rigorously and scientifically, is, of course, a different matter entirely.

You've written a popular book called How Intelligence Happens, and I must confess to you that before I began to read it I was enormously sceptical.

My thinking was something like this: You can't possibly just assess somebody with a number, you can't put a label on somebody like that. What's the scientific basis for that? And isn't this all just hand-waving anyway, because we don't really know what we're talking about. After all, all too often we can't even figure out a way of successfully navigating robots around a crowded room, let alone talking about these extremely complicated, sophisticated cognitive processes.

So I began with, I think it is fair to say, a fairly closed mind. But when I started reading, I quickly found myself engrossed in the story because, as a scientist, you naturally stay away from the sorts of unfounded hyped-up claims I had feared.

You begin the book by pointing out a particular piece of statistical evidence: a correlation that has been known to exist for a very long time that is scientifically rigorous and begs for some sort of scientific explanation.

JD: Yes. Let me first say something about this question of measuring people with a single number. Right from the very beginning of scientific study of cognitive abilities, which goes back now just over a hundred years, it's been perfectly clear that you cannot assess somebody's abilities, and certainly not their worth, through a single number.

In fact, there are an infinite number of things that are important to people and we value them for so many different characteristics: their honesty, their good humour, their ability at math, their ability to play football and so on—an infinite number of different things. That is the truth; and if you think you can explain everything important about somebody with one test, it's obviously doomed. As far as I know, nobody has ever thought that, although many people are criticized for thinking it.

However, what started with the work of the British psychologist Charles Spearman in the early part of the 20th century was the study of something much more specific than that. That was, as you say, an empirical discovery drawn from experiments on measuring people's ability to do things; and putting forward a theory to explain it.

The basic discovery is that if you give people a wide set of different things to do—they may be tests of memory, of problem-solving, of identifying briefly-flashed stimuli, of deciding which two weights is heavier, and many, many different things—you find that, overall, there is a tendency for all of these things to be positively correlated with one another.

This means that, on the average, to some extent it's true that a person who does well on one thing is also more likely to do well on other things. And the extent to which that's true is a measurable quantity.

There were several possible theoretical explanations for this, but Spearman proposed that perhaps there was something in the mind or in the brain—though he wasn't in a position to think in terms of the brain then—which contributed, to a degree, to success in all sorts of different activities that you undertook. He called it the "g factor" or "general factor". He intentionally kept it abstract like that because he had no idea what in the mind might produce this effect.

Since then, I suppose, a large part of the study of differences between people has been an attempt to understand what it is, in terms of either cognitive information processing or brain processes, that could be the underlying explanation for Spearman's idea of a g factor.

But it's worth remembering that it is a theory—there are other possible explanations of the data—and it's there to explain a particular pattern of results, which is this universal positive correlation between different things.

Questions for Discussion:

1. What do you have in mind when you describe someone as "intelligent"?

2. Do you think that intelligence can be developed or is something fixed that we are born with?

II. Trusting Your Gut

Intuitions and contradictions

HB: In my mind, there are two very different aspects of this. One is the phenomenon itself and the other is the possible explanation.

The first thing that really struck me in your book is that there is this concrete statistical result based upon a wide sample of individuals. We have a strong, empirically valid, correlation: people who are good at any one thing will also tend to be good at something else, which may be completely different.

It may be music or drawing, it may be reacting to stimuli, it may be solving puzzles or whatever, and of course the correlations are not one-to-one and there is a wide variety of human experiences. But this idea that you have a strong statistical piece of evidence is extremely suggestive that something very clear is going on.

And I was not aware of this at all. Maybe I'm the only one around who's not aware of it, but I suspect not; I suspect that my ignorance here is fairly typical.

Now, what the underpinning theoretical explanation might be to best explain this is, as you say, another matter. But I think it's really worth emphasizing that very few people would argue that this is a piece of statistical scientific evidence that calls out for some sort of explanation.

JD: Yes. I think that if you're going to study the mind and the brain in science you have to be doing it this way. You have to start the same way you'd do any other aspect of science, but it's more difficult with studying ourselves because we come with this advanced baggage of how we think about human nature. But I think if you're going to make progress in a science it has to be done this way: you have to start with observations, either quantitative or qualitative, and theoretical structures to explain those observations. And you must try not to step too far out of that to say, "How does this relate to how I used to think about the mind and the brain when I was a normal person rather than a scientist?" That often gets you into severe trouble.

Regarding what you say about your assumptions before you read the book, I think there's actually something very important there about our everyday thinking concerning ourselves. And that is that it's immensely flexible: people are perfectly happy to entertain flatly contradictory views without ever noticing that they are indeed contradictory.

In everyday life, for example, we all talk about equal opportunity. Americans, for example, regularly tell their children that each one can be president one day.

Meanwhile, it's very common to hear that everybody has their strengths, and that's a great thing. But then you go into the job interview next door and people are saying things like, "I really like that candidate, that's the smart one." Well, what do they think they mean by that? They don't believe this person will do well at the job because of being good at some one particular thing, but rather because they just somehow felt that they got more out of them intellectually.

I think people always have an intuition—which to some extent, at least, is true—that the same people tend to be able to flexibly address themselves to different sorts of problems. Meanwhile, very importantly, we have all sorts of more individual talents, skills, areas of expertise, which are also significant. I think people can point themselves either one way or another in everyday thinking.

But what's important about a scientific program is that for a given question, such as predicting how well a person will do in a new job when you're interviewing her, this is something that you can actually measure.

Questions for Discussion:

1. Is there a danger that scientists who are drawn to study intelligence have more pre-conceived, non-scientific ideas about the subject than others?

2. To what extent is your intuitive notion of intelligence related to "knowledge"?

III. Paradigmatic Examples

Some tests are more g-relevant than others

HB: Another point that struck me as particularly interesting, and somewhat counterintuitive, was that, while this g factor could be assessed by a wide statistical sampling of all sorts of different activities, there were some specific tests that, by themselves, could also provide a very good measure of the g factor. That was a surprising result, but it also gave me a clearer sense of the scientific structure and aspects of the original motivations behind intelligence testing and IQ testing.

In particular, there's this notion of g-saturation, the idea that you can take one or two tests and actually have some statistical measure of the g factor across a wide cross-section of the population.

JD: I agree that this is possibly the most important thing that's arisen out of the whole experience, the whole theoretical take on intelligence and the g factor. The fact that it then becomes an empirical measurement: what sorts of tasks or activities are most related to g?

And it turns out that by far the best tests are types of simple puzzles that you might find in a book of children's puzzles.

What does that mean to say they are the best tests? It means that when you measure how well a person does these puzzles, you get the best broad ability to predict how well they would do in all sorts of other things that they undertake. So whatever it is that these puzzles are measuring, it seems to be something really important about their mind and brain. It's like a flag stuck in the sand, if you like, saying, "*Look here!*" to the scientist.

If we can understand what's going on in these tests then we've made a big step forward in understanding what the g factor might actually be. Let's take a look at a couple of these tests.

The first one is what is called a matrix test, it's a typical, simple problem-solving activity of the sort you might find in a book of puzzles.

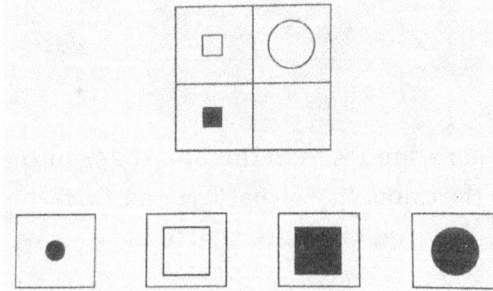

In this case you're supposed to decide which of these alternatives at the bottom correctly fills in the missing square at the top. It involves a certain amount of dividing the problem up into parts, thinking: *Well, I'll take into account the difference in shape between left and right, and I'll take into account the difference in size...* Eventually once you've got all the bits of the puzzle worked out in your mind then you can come to the correct conclusion that the answer is going to be the large circle that represents the last of the four options in the bottom right-hand corner.

This is a kind of mid-level difficulty of a typical g-related test. It's a problem that, I would say, roughly speaking, half the people will get right and half the people will get wrong.

Here's another example:

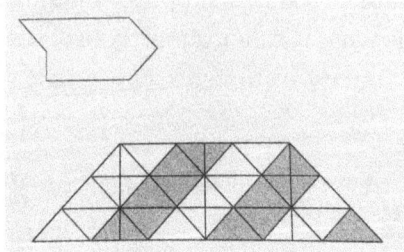

In this case you're supposed to find the shape at the top hidden somewhere in the camouflaging background. Once more, this is the sort of activity that's quite well-related to the g factor.

Here's one more:

1, 2, 5, 26, ?

You are presented with a number series where you start off with 1, 2, 5, and 26, and then you're supposed to work out what the next number would be.

In all of these examples, the critical cognitive step is to find the sub-problem— the part of the thing that allows you to make a signifi-cant step forward towards the solution.

Let's look at the series example. I've solved this problem before so it now seems obvious to me that there's one place where you're going to get the solution from. It's going to be very hard to work out anything about the series from 1 and 2, because there are so many relations between 1 and 2. And going from 2 to 5 is still not at all obvious for many of the same reasons, but moving from 5 to 26 you think to yourself, "How on earth am I going to get from 5 to 26?"

At that point, squaring 5 is likely to come to mind, and now you've identified something that's a big clue in the puzzle. Once you've squared the 5 and added 1 then it's pretty easy to go back to the rest of the terms in the series and see that that's the correct rule.

HB: So it's this idea of breaking things down, imposing a plan or a goal, having a structure to one's thought processes. That's what's so vital to success in all of these types of puzzles.

JD: That's right. My hypothesis—as I would call it at this stage, since it's certainly not accepted scientific fact—is that the process of finding a useful structure within complex problems and dividing them into solvable parts is the absolutely critical thing that g is concerned with.

If that's true, then it relates to many other aspects of the study of intelligence in an interesting way. For example, I mentioned earlier the activity from the mid 1950s onwards of trying to make computers think like people: the field of artificial intelligence.

Right from the start there was something very interesting that became apparent. It turns out that much of what we can do, like simply looking around a room and understanding what objects are where, computers are still absolutely hopeless at. What we do effortlessly computers are still nowhere near able to do.

But much of what we think of as 'intelligent' thinking and reasoning, computers were very good at right from the start. The first serious artificial intelligence enterprises were cracking problems like proving theorems in formal logic. This was the work of Newell and Simon and others in the 1950s. This is the sort of problem that we look at and think, "*Wow. That's something really brainy, very challenging*". And that turned out, for some reason, to be one of the easiest things to adapt computer thinking to.

In my opinion, this is also a major clue: the whole secret of how computers think is to begin with a complex problem and identify within it some smaller sub-problem that it knows the solution to.

Then it concatenates these into a whole series, like the steps of a mathematical proof or something. This is what computers like to do and I think this is really quite close to the heart of what's measured in tests related to g.

Questions for Discussion:

1. How would you describe the critical cognitive step required to solve the embedded shape puzzle in this chapter?

2. How might the field of artificial intelligence have been significantly influenced by our own biases of what "intelligence" is?

IV. Different Types of Knowledge
Crystallized vs. fluid

HB: One of the things I'd wanted to talk to you about is how these sorts of tests relate to IQ tests, because I think most people are familiar with the notion of IQ tests, which can be quite controversial. Many have pointed out that such tests may be culturally dependent, language dependent and so on. In short, they may be knowledge dependent.

From reading your book, I understand that intelligence tests in the psychological community seem to break down into measuring two different types of knowledge or awareness:

The first is something called "crystallized knowledge": things that actually do depend on what we've learned before.

The other is something called "fluid knowledge": the capability of solving new problems without necessarily a huge amount of previous knowledge or techniques, that we're able to somehow form a way of finding the solution using our mental processes.

And the sorts of tests that we've just been discussing here that are such great independent measures of Spearman's g factor are tests of fluid intelligence, right?

JD: Yes, absolutely. Again, this is a distinction that I think we're not always aware of in our everyday culture, although once it's pointed out it's obvious.

There are many things that we tend to call intelligent that are simply a reflection of the kind of education that we've had. So, for example, people tend to be impressed with a person who uses long words or is able to construct elaborate sentences or make mathematical proofs. But to a large extent, this has to do with what you were taught rather than the sort of person that you are.

Then there's the other side, which, as you say, is trying to test fluid intelligence. In this case, a serious attempt is made to minimize the amount of knowledge that would be needed. These naturally vary in that respect: obviously if you haven't learned anything about squares of numbers you're not going to solve the problem of the mathematical series.

HB: Sure. But that's sufficiently general that most people would have that awareness.

JD: That's right. In the fluid intelligence tests, the attempt is to make sure that the basic knowledge would be available for all the people in the community you're trying to test. Of course, that may be learned knowledge. It may be that if you went to a completely different culture then the test would no longer be applicable. But the attempt is to try to measure something about the basic brain function rather than about the particular knowledge that has been put into it by a person's experiences.

We can argue about the extent to which that's completely successful, but certainly it's true that fluid intelligence tests have very different properties from crystallized knowledge intelligence tests. The most salient case is what happens in old age. In old age, sadly, fluid intelligence falls off catastrophically. It probably peaks somewhere in your teens and then changes rather slowly into your fifties and then begins to really drop off precipitously.

Whereas when it comes to crystallized intelligence tests, such as tests of vocabulary or one's ability to write a nicely integrated piece of prose, there's very little evidence that, until you really start to advance substantially in age, that sort of thing is going to decline.

The basic idea, then, which has been around since the 60s and probably before, is that knowledge, once you've acquired it, is fairly solid. It takes substantial degenerative changes in your brain for it to disappear.

Questions for Discussion:

1. To what extent is it possible to objectively distinguish between "crystallized" and "fluid" intelligence?

2. Why do you suppose that "fluid" intelligence drops off so sharply in middle age?

V. Another Correlation

The power of organization

HB: In the mathematical science community there is this sense that if one doesn't make a great breakthrough by the time of twenty five or thirty then it will probably never happen. That's a bit of an exaggeration, and there have been many examples of people who've made significant contributions as they proceed in life, but for the most part there seems to be something to it, statistically: many of the greatest and most revolutionary ideas in the mathematical sciences came from people who were quite young.

Whereas when you contrast this with the work of a historian or the work of a judge or the work of people who are distilling vast sums of knowledge and utilizing their repertoire of experiences, things tend to go in the other direction. Often someone's great philosophical or historical life's work will be written when they are in their 50s or their 60s.

These ideas seem very much in keeping with what you've been describing here: the mathematician, say, would be relying more on this notion of "fluid" intelligence whereas the philosopher, the judge and the historian would be relying more on this collection of experiences represented by what you called "crystallized" intelligence.

JD: Yes, of the ability to integrate a large body of knowledge, which gradually is built up through your lifetime.

HB: Exactly.

JD: I think that's almost certainly correct. Charles Spearman himself, incidentally, was very concerned about this puzzle. He seemed to have invented this method for measuring the g factor that was so important in a person's life. Pretty soon he discovered that this g would be at a maximum in one's 20s and then he started to question why it is that our governments are run by 70-year-olds and all major activities tend to have their organizational aspects overseen by older people.

I guess it's another proof of what we talked about at the beginning: that g, I think, is something important and very interesting to understand at the level of brain systems. It's only one thing about you and by no means it's to say that it is the best predictor of everything that you may undertake.

HB: OK, but it does exist; this statistical correlation represented by g is real and we've got various ways of measuring it reasonably precisely. And since we all believe that thought processes originate in the brain, if we want to study why people are better at certain cognitive tasks than others then we've got to look at the brain. The question is, where, exactly, is it in the brain?

JD: Yes, well, let me step back and talk about the layout of the brain and what the different parts are good for.

Much of our early knowledge on this comes from the 19th century onwards, from the consequences of damage to particular brain regions in patients with strokes or tumours or other neurological conditions, and the picture that you see is a fascinating patchwork of functionally specialized regions doing different things.

For example, most people probably know that visual information hits your eye and then it's passed along nerve fibres to the back of your brain in the occipital lobe where it's processed to give an understanding of the layout of the visual world around you.

If a stroke patient has suffered damage right at the occipital lobe, then he can either be rendered completely blind if it's too close to the beginning of the system, or he might have an inability to recognize faces or to navigate his way around the visual environment—there are many separate brain modules, if you like, doing specialized things.

Similarly, it's known that a stroke in the left hemisphere can mean a person can lose the ability to understand speech or to generate speech or the combination of the two. Sometimes quite fascinating tiny little islands of inability arise, such as the inability to read simple function words like 'a' or 'but' or 'the', and yet those afflicted can still read a word like 'chair' or 'crocodile' or 'hippopotamus'.

So almost every aspect of behaviour you can think of: memory, control of the body, reaching out to a particular part of space, being aware of things on one side of space, can often be damaged in a brain injury.

All of these things suggest a picture of substantial modularity of a complex organ in the brain with different parts which are responsible for different parts of our mental lives. In that context, what could be the basis for Spearman's idea of g? And let's remember again that g is just a theory: there are other ways to explain the data.

From the study of brain damage and then later from the study of functional brain imaging with PET scans and functional MRI, an answer started to emerge. And it goes back to these puzzles that we were looking at before. We can ask, "What is it in the brain that is active while people solve puzzles of this sort?"

And it turns out that the answer is somewhat complicated but also surprisingly simple: it's far from the whole brain that is involved when you're doing tasks of this sort. Instead, it's a very particular network in the frontal lobes—right behind your forehead—and the parietal lobe, which is sort of on top and to the rear, if you like.

Very interestingly, if you gave many different sorts of tasks—tests of memory or language, or even identifying faces or holding something in short term memory—the same network tends to be a part of the brain's response.

Often, all the individual modules that are important for a particular task, such as the language module, are active too; but a significant part of the brain's response is this core network that seems to be related to intelligence or g.

So whatever g is, it's something that's important in organizing many different sorts of activity, which is just what you should think if it's going to explain the g factor. And much of what we're doing now is trying to understand what's going on inside these regions of the brain as problems are solved.

HB: When you first described these ideas in your book, you talked about some specific experiments that you were doing.

You mentioned how your colleague was conducting tests on patients very similar to the ones you showed before with the camouflaged shapes, while you were doing other experiments.

In yours, subjects would have to identify specific sounds coming into either their right or left ear after hearing one of two types of tones. If they heard one type of tone they would have to focus on what was coming into their left ear, and if they heard the other type of tone, they should focus on the sounds coming into their right ear.

And after a while you became struck by the fact that quite frequently the same people who had difficulty with the task of finding the shapes also had difficulty with identifying the specific sounds with the right ears.

I was particularly struck by the fact that the people that were having difficulty with your experiment fully comprehended the basic framework of the test. You would ask them if they understood what they were supposed to do and they would reply *"This is what I'm supposed to do"*, but somehow they didn't actually do it. This was somewhat mystifying to me.

JD: Yes. This was possibly one of my first exposures to the power of g, that is, the power of a positive correlation of the same people tending to stumble and have difficulty in apparently quite different situations.

As you say, one of the tests we were using, with my colleague and friend Frank McKenna, was finding those figures hidden in camouflaging backgrounds.

Some people have immense trouble with doing that task. Some people can sit staring into space for literally minutes without making any progress. We've now studied this sort of thing for over 30 years so we know quite a bit about this phenomenon.

It doesn't actually matter what the task is. What matters is that the task had a number of different parts to it and you would give verbal instructions explaining what it was that the person was supposed to do. In this case it happened to be, as you say, listening to tones, deciding whether to pay attention to speech in the left or the right ear, listening to what they were told, repeating some of the words, shifting ears.

It was a task with multiple parts to it and what happened was that occasionally somebody would clearly have verbally understood everything that you told them to do, they could repeat the rules back to you perfectly, and then they would do the whole task completely wrong—part of it was just left out as if it didn't exist. It was very striking that this whole pattern of behaviour would continue, not with any sense of disturbance, but just as if it somehow hadn't clicked that this wasn't what they said they were going to do all.

As it turned out, it was essentially the same people who had trouble with the camouflaged figures as had trouble with organizing their behaviour to this new set of relatively complex instructions. And indeed, these were people who would do badly on standard tests of fluid intelligence.

Once again, this phenomenon seems to be a strong flag in the sand saying, Look at this, look at this failure to do what you said you were going to do, because it's strongly related to whatever psychological or brain processes are measured in g.

This immediately caught my attention because in the literature on people with brain damage there's a type of patient who is known to act like this in a more extreme way, even in very simple tasks.

So here's a classic example. The experimenter says, "When the light goes on I want you to lift your hand up". The light switches on, the patient thinks, I should lift up my hand. Yet they sit there doing absolutely nothing.

Now obviously they can lift up their hands. In fact, if the experimenter then says, "Look here: I switched the light on, you should lift up your hand", then they do it. But somehow the knowledge of what was supposed to be done and the conditions of the light coming on wasn't enough to draw out the right behaviour.

Their performance was the same from what I saw before in tests with patients with damage to a particular part of the brain, the frontal lobes. And this, in a way, led into the brain-imaging experiments that later highlighted this core intelligence network I was speaking about a moment ago.

Questions for Discussion:

1. *How do we know when we have identified a specific "network" in the brain? What does that mean, exactly?*

2. *Might there be other reasons to explain the typical difference in age between those who create impactful work in mathematics and history other than by invoking notions of "fluid" and "crystallized" intelligence?*

VI. Selecting Solutions
Finding focus

HB: So it's as if these people have a difficulty in imposing their will somehow. They may understand something in theory but somehow they don't seem to get to the point where they say to themselves, "Now I'm going to solve this problem using the information that's available to me."

It's one thing to not be able to visualize a particular shape in a larger series of shapes, but it seems a different sort of situation altogether when somehow there's a missing link between, "I understand what I'm supposed to be doing about tones and listening to the sounds coming into the headphones" and then going ahead and actually implementing those instructions.

JD: Right—people often wonder how much this has to do simply with motivation. You can do experiments where you say "*OK, if you get this one right you are going to get $500.*" I've never tried that experiment, partly because I am pretty convinced that it doesn't work.

Often you see people who really are struggling: they want to do what you're asking them to do but still they just cannot get it all organized in their minds. My take on it is that it's much more to do with the ability to focus on the exact part of this complex situation that is important right now. We talked about that a little bit in the case of solving the number series, where you find something and focus on it: you use the key to unlock things.

Even when you're doing much less of a problem-solving type of task—simply following instructions, for example—I think this constantly happens. If you give someone instructions for a task and then say, "Now do it", it all just becomes a sort of complex blur. What you need to do is pick out what exactly matters and produce a little episode of attention to that, solve that part of the problem and then move on.

I think this is really what was happening in my test. For example, for those people who ignored the tones as they came in, they had this general sense of, There are words, there are tones, there's me sitting there: it's a whole complicated situation.

It didn't happen that when the tone sounded their brain focused on that and they thought, "*OK, tone: that one is low, it means the right ear.*" Of course, when we do that sort of thing we have the experience of an act of will, if you like, but I don't think it's the simple motivational aspect of it that's important. I think it's much more the organizational aspect of it.

HB: So somehow the key is to find the way to focus on "the right things".

JD: Yes: selection. Focus is really the same thing as selection, if you go into it. It means that you take some specific part of the problem and not the rest. Interestingly, I think this is also very closely related to what many people think of as the heart of what's special about the human mind: the power of abstraction, the idea of being able to think abstractly.

Often we tend to think of abstraction in, if you like, rather abstract terms: we think that abstract thinkers are philosophers or those who can think complex problems. I think there's probably a simpler way to think about abstraction as the ability to see the common important thread between a great many instances that are different from one another.

This is what we mean by abstraction. An instance of justice occurs, for example, no matter whether it happened in the court or on the playground. You can see we're now very close to the idea of attention or focus: we're picking up just one critical aspect of a situation and throwing away all the other things that differ between them.

I think that probably this brain network we associate with g is very much related to this heart of human thinking: the ability to abstract out just which aspect of a situation is critical for the current moment.

HB: There's a wonderful example you give in your book when you talk about the problem—I think it's called "the mutilated chessboard problem"—of how to tile a chessboard with two pieces that are removed.

One imagines a chessboard with the two light colour squares in the corners removed, and the question is whether or not it is possible to completely cover this modified board with dominoes, each one being the length of two squares of the chessboard.

You mentioned that intelligent, reasonably educated individuals might take an enormous amount of time to try, vainly, to develop a solution.

But when you focus on one key aspect, namely that any particular tile that is two squares long necessarily has to cover at least one pink and one black square, you immediately recognize that the problem is impossible for a chessboard with two white pieces removed, simply because now you have two more black squares than pink ones and there is no possible way to cover the "extra" black squares.

That simple little bit of information—as you put it, the key piece of information—is vital.

And I'm guessing that what you mean by being able to focus is linked to being able to concentrate on matters in such a way as to be able to find that key that unlocks the problem.

It seems like the frontal-lobe patients that you were talking about before are almost the opposite extreme: they have great difficulty selecting the right piece of information, they are disorganized, they are distracted.

Another thing that struck me as particularly interesting about this example is that this is not necessarily the sort of thing that is normally meant by intelligence. We normally associate intelligence with someone who can come up with the right answer straight away.

JD: Yes, what you say about speed is very interesting.

One very well-established camp in thinking about the g factor holds that it is very much to do with mental speed. What is the evidence for that? Well, the evidence is that if you measure how quickly people can press a button in response to a light, shall we say, then it does correlate with other things including tests like the puzzles we've been looking at here.

But of course we've already established that it's a universal law: all tests positively correlate. So I am not particularly keen on a simple speed explanation for the obvious reason that if you measure how fast a person can press a button in response to a light then, yes, it has some positive correlation to other things, but the correlation is not that high.

Whereas if you use the specific sorts of tests that we discussed earlier, then the correlation is much higher. To me, that says you're getting much closer to the critical cognitive or brain processes by looking at how these complex problems are solved and likely divided into parts, instead of looking at simple mental speed.

That being said, of course it is true that how quickly you can do something is related to g. There's no doubt about that. And in my opinion, the more complicated the task, the better that correlation tends to get. Again, I think this probably reflects the same thing: to do complicated things fast, it usually means you have to be focusing on the exact thing you are doing right now as opposed to what you were doing a few hundred milliseconds before. My suspicion is that this is really the core of why speed is often thought of as so closely related to what we think of as "smartness".

Questions for Discussion:

1. Can you imagine specific examples of high intelligence that have nothing to do with speed?

2. To what extent can you think that focus or selection can be learned? Does doing lots of puzzles make it easier to focus and find "the key"?

VII. Looking Inside

Harnessing modern technology

HB: So let me pause for a moment and see if I can recap where we are. Over a hundred years ago, based upon a wide-ranging series of tests, Spearman notices a statistical correlation between people's measure of success across a wide variety of different activities. From there, we come to appreciate that some tests are better indicators than others of this statistical link or "g factor".

Then you and your colleague do some other experiments, not necessarily thinking about g at all; and through these experiences you develop the idea that you may have discovered some deeper understanding of what, physiologically speaking, might be fundamental to g: that the same people who are performing badly on standard g tests seem to be acting in very similar ways as people whom you've read about in the frontal-lobe literature. This leads you to suggest that g is, neurophysiologically, somehow linked to the frontal lobes.

So let's examine that in more detail now. What's really going on in the brain when we're engaged in these tasks? What's the diagnostic equipment telling us? Combining your intuition, theoretical models and empirical studies, what do you think is going on, exactly?

JD: This whole thing, of course, is progressing at lightning speed— well, it seems like lightning to me, though I would say that the relevant time intervals are decades and half-decades, rather than minutes.

When we started this line of thinking in the early 1980s, our evidence about assignments of function to different parts of the brain came largely from brain damage, which is, of course, extremely coarse.

Why? Animal experiments give you better data but human experiments are very coarse because the damage that you end up with varies randomly from one person to another. It rarely respects the exact boundaries of a particular parcel of the brain that you believe to be functionally important.

So it's hard to get anything other than a very, very blurry view, and that's why in the 80s we were thinking in general terms about the frontal lobe. The transformation that began in the late 80s has been an absolutely massive worldwide research enterprise: the invention of methods to use PET scanning and later fMRI to measure which part of the brain is most active while different types of activity are undertaken.

HB: How do these scanning methods work, exactly?

JD: In both of those cases rather indirectly. What we are actually interested in is the electrical communication between these tiny neurons. We probably have about 10 billion neurons in our brain, so you get a sense of how small each one is. Miraculously, they are able to extend: they have long lengths and can speak to one another despite the fact that they can't be seen without a microscope.

The individual neuron is the entity that you'd like to know about, it's the basic computational unit of the brain, if you like.

But in functional brain imaging in its present state, all we can measure is the average activity of millions of these neurons in a little bit of the brain about the size of a peppercorn. It's like listening to the whole volume of noise coming out of a city. Luckily it's not quite as bad as that because nearby neurons tend to be doing somewhat similar things, so you can get some sense of the signal, but you can imagine how crude it is.

Unfortunately, too, the neuron is communicating on a millisecond timescale with tiny little pulses of activity, but what we can measure with functional MRI is changes of blood flow caused by a bit of the brain becoming active and requiring more oxygen, which, in turn, changes the blood supply.

And that takes place over a course of seconds, so it's thousands of times slower than the timescales that we want.

HB: So what I'm looking at when I look at these particular scans is actually the increased blood flow that is indicative of energetic activity corresponding to certain areas of the brain where more blood is flowing in.

JD: Yes.

HB: So on the one hand it's terribly exciting because until recently it's been impossible to scientifically measure in real time, or close to real time, what people are actually thinking.

On the other hand, it's a good deal removed from what we'd ideally like, because we actually want to be directly measuring the neurophysiological firings that are going on.

JD: Yes, the communication of individual neurons and patterns of neurons, population of neurons.

HB: And PET scans effectively work the same way as functional MRIs, right?

JD: Yes. In fact, they work at even much lower timescales. Functional MRI was a substantial step forward in terms of temporal precision. But we are still very limited.

So to go back to my story, what we see with fMRI, despite its limitations, is much more refined than what we could ever have understood by looking at brain damage.

We're now not talking about the whole frontal lobe by any means, but rather a tight little network within the frontal lobe that we can see with fMRI, and similarly for the parietal lobe. We never could have focused so precisely on these areas from examining patients with brain damage.

This method is useful for targeting attention on the exact bit of the brain that we're interested in, but it's still very limited in terms of how much you can peer inside and computationally assess actual thought on the millisecond or ten millisecond timescale.

It can tell us some things but there is a lot that it can't do. I'm sure that fifty years from now this method will not be used. Instead, we'll have invented ways to get highly spatially accurate and temporarily close measurements of actual electrical events in the brain that at the moment we simply don't have.

HB: Let me just interject with a small pragmatic question. I've had an MRI on my knee, and it was extremely noisy. I could imagine that if somebody was told to think about a particular set of thought processes and lie very, very still, it might be quite challenging to do an experiment that can precisely measure brain activity while all this noise is going on. Is that a problem? Are there others?

JD: It is challenging but the challenges aren't necessarily the ones that would immediately come to mind. You do have to get the person to lie still and they generally can.

They have a screen in front of their eyes that is mounted inside the MRI scanner and events occur on it: they are given some problem and have to press some button corresponding to a particular solution. That all works reasonably well. The noise is loud, but usually people are equipped with earplugs so the sound is not too bad.

The real problem for me and many others is that you're lying there in a rather dark environment surrounded by some noise. It's like taking babies out in the car: pretty soon you're struggling to stay awake. I think it's certainly true that the results of this method are heavily influenced by people dozing off or almost dozing off.

The other problem is that it's rather a confined environment, so there are significant numbers of people who as you slide them in say, "*I want to get out*". The first time you go in you feel a slightly claustrophobic panicky sense that things are out of your control. But very quickly you adapt and then you're bored stiff and, as I say, dozing off.

Question for Discussion:

1. Would you be willing to volunteer for one of John's experiments in an fMRI machine?

VIII. Implications
Future possibilities

HB: What are the implications of your findings in terms of future work: what you'd like to test and how you'd like to develop your theories in terms of the physiology of the brain.

JD: Well, this is all very much ongoing of course. I would say we're very much closer to the beginning than to the end of the enterprise, but let's summarize where we've got to and what we've talked about so far.

We've identified a particular network of structures in the brain. We believe that these are important in the g factor, assuming that's the right explanation of Spearman's data. We know they're important in organizing the brain's response to all sorts of different cognitive challenges or tasks because we see them active so frequently. And we strongly suspect—although it is just my hypothesis and by no means universally accepted—that what we should be looking at is the ability to decompose complex problems into a series of focused processing or attentional episodes that allow effective solutions to be found and the structure of goal-directed behaviour to be created.

These are the leads we've got from what we can do at this level. Now what we would like to do is peer inside these regions of the brain and find what neurons are actually doing and how patterns of activity evolve as problems are solved or as an attentional episode is created. We're very much in the early days about this.

To a degree you can do it with fMRI because, crude as it is, if you look at the exact pattern of activity within one of these regions—for example if they are doing a task where they have to discriminate between the letters X and O—you could look at the exact pattern of activity for Xs and the exact patterns for Os.

Remarkably enough, crude though the picture is, there's enough difference sometimes to be able to read what it is that the person is actually thinking about. So it is kind of a first step of reading somebody else's mind with a machine, coarse though it is.

HB: You can detect differences at that level? That's remarkable.

JD: Yes. These experiments have been going on for five to ten years now and I'd say we're still somewhat finding our feet with the exact methods that work best. But to a degree, it works surprisingly well, especially in the visual system. You can get quite detailed information about what sort of image a person is looking at.

That is one method by which we can investigate further. Suppose I ask you to focus your attention on two different things. On one trial I'll ask you to pay attention to a cup and on the other trial I'll ask you to pay attention to a keyboard.

Then we can see that in this part of the frontal lobe the representations, to a degree, follow that. It seems that the neurons are able to take up the information that you're paying attention to right now, just as you would want if the system is constructing this little micro fragment of cognition, if you like.

If this is true, it's very interesting because it's not at all in accordance with the established neurophysiological rule of how the brain works, where people tend to assume that each particular neuron does a particular thing.

For example, it recognizes a pair of staring eyes if it's part of social processing or, as in the hippocampus of rats, a single neuron will turn on when the rat is in a particular part of the environment and the neuron will be off all the rest of the time as if the structure is building a map useful for navigation. That's the system that's being worked out in quite a bit of detail in rats.

But what we seem to be imagining is true in this core intelligence network, if you like, is something quite different from that. Here neurons are extremely flexible in their properties and instead of just doing one thing all the time—being specialists—they are now generalists and will pick up the information that you need to be thinking about right now, while at the same time filtering out all the other distracting parts of the cognitive blur around you in the world that would stop you focusing on what really matters.

HB: It's amazing to imagine what sort of flexibility that might entail on the neuronal level for neurons to be able to do all sorts of different things.

JD: Indeed, that's also a fascinating question. Let's just think about a neuron in the occipital lobe, in the visual system. Predominantly what that neuron gets is visual information. You can trace visual information and its connections back to the retina, and then, through multiple steps of visual information, to what arrives at those neurons.

The frontal lobe is connected, directly or indirectly, very widely to most other bits of the brain and certainly the cerebral cortex, so it's getting many, many different kinds of information. So these neurons in the frontal lobe communicate very widely with one another and it's very plausible to believe that each neuron actually has access to many different kinds of things, and therefore there's the potential for it to give many different sorts of activity patterns.

Then you need to think of a physiological mechanism by which its inputs are gated so that whatever matters right now drives the cells' activity and everything else is kept out. How that could happen is a fascinating question, but I'm very strongly persuaded that something of this sort must be going on.

Questions for Discussion:

1. What do you think John means, exactly, when he talks about "inputs being gated"?

2. Are there moral concerns with the development of increasingly accurate brain-scanning technologies?

IX. Assessing the Landscape
Reactions and speculations

HB: Let me return to a couple of points now. You were very clear throughout this conversation that what we were talking about was your theory: it's your particular way of interpreting Spearman's data and your explanation for g: what it means, where it exists and how we might move forwards to more accurately probe it.

Presumably there are other people who think differently. What are some of the other views, the other interpretations of what's actually gone on. What are some of the criticisms that others will level against you?

JD: Sure. But first, you talk about this as my theory, so let me duly register my British embarrassment about that. It's probably true that only a few people would agree with the integrated story that I've told here, but there are various sub-parts that quite a number of scientists around the world would hold similar views on.

That said, as you say, there are a lot of alternative takes on many aspects of this problem. I think probably the most widespread view in experimental psychology is that Spearman's idea of g is highly misleading, that positive correlation should be explained in a different way.

It was first well articulated by another British psychologist, Godfrey Thomson, in the early part of the 20th century just after Spearman's work, but I think it was implicitly believed or explicitly stated by most people in the field.

The idea here is that there is no common reason for positive correlations: that any two tests correlate simply because they are likely to have some aspects in common.

We talked before about the brain being filled with different functional modules. So the idea would be that if you take any two tests they are likely to have a couple of modules in common, so they'll show some positive correlation. But there's no universal explanation for a broad pattern of positive correlations.

HB: So in Spearman's words, when he talks about his "g" and his "s" as the two units to be able to indicate whether or not somebody would be good at any particular task, the idea would be that there is no g because there's just s. And g just emerges as this average of all these s's for any particular task that you have?

JD: Yes. So, the s's in modern brain terms you might think of as being the functions of all these many different brain modules. That, I think, is still probably, explicitly or implicitly, the most popular view, and it's easy to prove that if you simply look at data on correlations between people's performance on different tasks the two theories are completely indistinguishable. They can always be moved around to explain one another's data.

In my opinion, the brain-imaging data is probably the strongest argument against that, because if you believed that what was really going on in problem solving tests like these involved all the different modules—

HB: You'd see these lots of different modules firing at different times.

JD: Yes; and to see one core system that seems to be specifically important is very much not what you would have expected from that point of view.

So now the debate changes, it moves to a new level. If we accept that this core system is somehow the most important thing, can it, in its turn, be divided up into separate functional parts? If so, then g would have several separable components to it.

And I think, at least physiologically, the answer to that is almost certainly going to be "yes", because within this network are included not only some really very anatomically different structures, but also very different connections to other parts of the brain. So it's extremely unlikely that all the different parts of this network are doing exactly the same thing.

But how might we break the network up into its components and see how they interact with one another to build attentional episodes? That is presently unknown and is a major question.

HB: But that objection strikes me as more of a refinement of the idea of g, rather than an argument in kind against it as the "multiple 's' argument" would be.

JD: Well, one man's refinement is another man's "novel, individual contribution". I agree with you, but many wouldn't.

Anyway, I certainly think this is going to be one of the major topics for expanding this line of work further.

Another thing that we haven't talked about, but is worth emphasizing, is that if we want to definitively answer questions like this, fMRI is a very crude technique for doing it. What would be much better would be to look at the actual individual neurons within each structure at the timescale I work in and then perhaps work out how a pattern in one part of the network then feeds on to influence and create a pattern in the next part.

For this experiment you need electrodes in the brain that can actually record neural activity. This can be done in human patients, usually for epilepsy patients, who are being explored to find the epileptic focus before surgery.

One can have many electrodes inserted in the brain: it's all completely painless because there are no pain receptors in the brain.

That would give a window on what neurons are doing as the person undertakes goal-directed behaviour. Similar sorts of experiments are being done with rats and they give a much more detailed picture of what I was talking about.

We can see that in the regional lateral frontal cortex, as their behaviour unfolds millisecond by millisecond, you can see these tightly focused patterns of activity that pick out the particular thing that the animal has to be processing right now.

HB: So we need better technology, we need better processing power, and we need those physicists and mathematicians to work harder to give us better diagnostic tools.

JD: Yes. I'm sure it will happen that somehow, either through fMRI or something else, people invent a method to directly measure what we need, which is the actual electrical activity of neurons—looking at the synapses as neurons communicate with one another. That will give you nice spatial precision of where you're looking in the brain. And when that happens our whole topic is going to be transformed.

At the moment we're frantically trying to see through a very blurred lens what's going on in the brain. But in the future we are going to have a beautiful lens and we're just going to have this absolute flood of data resulting from, as I say, literally billions of processing units each communicating through thousands of connections with other ones.

At that point the problem is going to be trying to condense this data into some form of comprehensible framework. This is the adventure of science. I think we have essentially no idea of how that enterprise will unfold, but I am quite sure that, one way or another, it is what we will be doing fifty years from now.

Questions for Discussion:

1. Why does Howard specifically mention "physicists and mathematicians" when discussing the need for better tools for neuroscience?

2. How might some current techniques for handling "big data" be one day applied to improved brain-scanning technologies?

X. Pure and Applied

Bringing it all home

HB: We didn't talk about this when we were discussing things before, but to me one of the most intriguing aspects of your early work with those experiments with the tones in their ears and all that, was that many of those individuals who were having difficulty focusing could actually be brought around to do the required tasks with enough work and patience.

JD: Yes.

HB: That seems a very hopeful sign. It makes one speculate that, if this g factor is, as we expect, some sort of sign of intelligence, then we might be able to increase our intelligence to some extent.

Maybe we shouldn't even use the word "intelligence", perhaps that's too loaded a word. But it seems to me that we might be able to improve our cognitive abilities in these standard tests of fluid intelligence, and it might thus be able to improve our own g factor.

Do you think that's possible?

JD: I'm certain that it's possible and I'm equally certain that we have very little idea at present how to do it.

Why am I certain it's possible? Because among the many things that have raised a downright antagonism and dislike to the study of g is the idea of its genetic component. Most experiments show that, indeed, there is a substantial genetic component. We are far from understanding the particular genes involved, but it's pretty clear that it has a strong genetic component. Everybody seems to be very interested in this: that's where attention is focused.

To me it's interesting but it's not the most interesting thing. The most interesting thing is that no matter what experiment you look at, there's always a strong environmental component as well. That means that somewhere in a person's life, stuff happened that really affected their scores on tests like this.

And since the scores are so predictive of how well you solve problems in your everyday life, that implies that some kinds of experience do enable a person to learn better to solve the problems that they care about.

And that is very big news: if you could only find a way to do it.

The law of research in this field is simple and depressing. It's very easy to train people to do a particular thing—including as you say, these experiments on attention switching. You can always persuade someone to switch his attention effectively with enough focused pointing out what he should be doing in the task at hand. But that's of no interest, of course. What's of interest, is getting it to generalize about something else in his life that he actually cares about. That's what's really difficult.

So training is very effective in the context of the exact thing that's trained, but developing broader generalization for other problem-solving is absolute murder. And though there are some hints of what things work, that's progressed very slowly. People have been extremely interested in the idea of training yourself to be brainier for a century or more. An enormous amount of work has gone into it and yet very little has come out at the other end. Not none, but very little.

I would love to devote some of my remaining life in research to the question of how to harness the ideas we've been talking about with respect to g—dividing complex problems into parts and producing good, focused attention on the components—and develop methods to teach people (probably children, because their brains are by far the most plastic) how to employ what they've got more effectively.

It's very plausible to me that, indeed, there's a great deal of variation in the way that adults interact with children in training them to pick a part of the problem, stick with it until it's done, and then move on in a nice clear way. That seems like something our interactions do help to shape.

I would very much like to do some research on how we can get children to use their entire mental machinery more effectively, not just in training them to do particular things well, like jigsaw puzzles or something. I'm sure there is an answer out there. It's just proving damn difficult to find it.

HB: Well, that's part of the joy and thrill of science, I suppose.

JD: Indeed. And of course I love basic knowledge: I think human beings are essentially born to love basic knowledge for a reason I mentioned right at the beginning of our conversation: that understanding the world around us is what's made us what we are and we are born with a thirst for knowledge. I think the love of basic knowledge does drive a good deal of science and that's very good that it should do so.

But meanwhile, I think a very good piece of evidence that you're actually studying something useful and getting genuine understanding is if it can be used to solve problems in the broader world—in this case, helping people to use their brains more effectively. I would love to have the basic scientific, theoretical work that we've been doing turn into an application of that sort.

HB: Well, you've put us very much on the right track. Thank you very much for a fascinating and stimulating conversation. John; it's been a pleasure to talk with you.

JD: Thank you. It's been a pleasure for me as well.

Questions for Discussion:

1. What do you think John means when he talks about children's brains being "by far the most plastic"?

2. Would successful methodologies to enable people to develop transferable problem-solving techniques amount to an example of increasing our brain's plasticity? Discuss.

Continuing the Conversation

Readers are encouraged to read John's book, *How Intelligence Happens*, which goes into considerable additional detail about many of the issues discussed during this conversation.

Minds and Machines

A conversation with Miguel Nicolelis

Introduction

From Science Fiction to Science Fact

As any marketing professional will tell you, knowing how to properly brand an idea is essential for its success. Indeed, all too often, it can be even more important than the actual idea itself.

The field of phrenology is a good case in point. A deliberately constructed combination of two heavyweight Greek concepts—*phren* (mind) and *logos* (structured knowledge)—its very existence argued strongly for its inherent validity. After all, how could a field dedicated to amassing a structured knowledge of the mind be anything less than rigorously scientific?

Well, it turns out that it can. Phrenology had its day, of course. And to be fair, when developed by Franz Joseph Gall in the late 18th century, it represented a significant advance in our understanding, emphasizing as it did the importance of explaining mental states through the neurophysiology of the brain rather than through the previous window of religious or philosophical abstraction.

But by the time people started engaging in detailed measurements of skull sizes to determine which one of the 27 arbitrary so-called "brain organs" was most responsible for someone's personality, it was clear that the field had descended to the depths of pseudoscience from which it would never recover.

Yet aspects of its legacy persist. According to Duke University neuroscientist Miguel Nicolelis, Gall's general framework of dividing the brain into regions that control separate aspects of human behaviour, *"morphed into one of the key dogmas of twentieth-century neuroscience".*

A dogma, as it happens, that Prof. Nicolelis is firmly convinced is, if not flat-out wrong, at least deeply misleading.

What he can't accept is the pre-eminence that classical theories of the brain give to a sense of locality: that the brain is partitioned into strict divisions of different types of neurons responsible for different types of information processing. For him, the fundamental thinking unit isn't a neuron at all, but rather populations—clusters—of neurons that are themselves distributed throughout many different areas of the brain. And when the brain acts, he believes, it does so by integrating all these distributed clusters of neurons.

> *"It is like the population of neurons is voting at each moment in time and the real outcome depends on this voting. It doesn't depend on a specific class of cells or a cluster of neurons. It depends on this distributed representation."*

Well, people have argued about how the brain works for centuries. What separates Miguel from his many predecessors is that he has managed to construct a good many convincing experiments that back up his hypotheses.

For much of his research career, he has distributed detectors throughout the brains of rats, monkeys and humans to directly measure their electrical signals. This, in itself, is astounding enough.

But what seems straight out of the realm of science fiction is that he *also* managed to control these signals and relay them to machines to enable the animals to independently control them *just by thinking*. And not just once or twice, mind you: over and over again.

Once he managed to get rats to drink from mechanical arms merely using their brains, he upped the stakes by illustrating how monkeys could be trained to think their way to remotely controlling a computer cursor.

Still, scepticism remained. After all, both experiments were using brain waves that had originally come from signals related to their upper limbs—the rats had been first trained to press a bar with their forepaw, while the monkeys had been moving a joystick with their arms. Perhaps, people thought, it was impossible to fully harness lower limb signals.

Now, frankly, this seems a pretty curious objection to me. After all, once you get to the point where you can manipulate animal thoughts to regularly move robotic devices, it seems you've proven your point. But suffice it to say that I'm not in the field. And apparently, for those that were, this upper limb/lower limb distinction was a pretty big deal.

So off goes Miguel, determined once again to show the critics wrong. He trains a Rhesus monkey to walk upright on a treadmill and measures the monkey's brain waves. Then he sends these signals to a processor that controls a 100 kg automated robot, so that as long as the monkey is preoccupied with walking, those walking-oriented brain-wave signals are converted into propelling the robot into moving as well. Meanwhile while the real-time image of the robot moving is beamed back to a video screen directly in front of the monkey on his treadmill.

So the monkey, just like many of us in our health clubs, gets used to walking on a treadmill and watching a large video screen. Except weight loss here is not the issue, the goal is simply to keep the robot walking. Because the monkey quickly learns that as long as the video screen shows the robot walking, he regularly gets a shot of his favourite fruit juice.

So what happens?

Well, the treadmill is turned on. The monkey walks, the robot walks. Then the treadmill is turned off and the monkey, strapped to a device and unable to move except on the treadmill, is forced to stop walking as well. But the robot keeps walking. Why? Because even though he is no longer physically walking, the monkey has made the connection:

he knows that if he is still *thinking about walking* he can still make the robot move with those thoughts and still get his fruit juice.

Oh yes, I forgot one little thing: Miguel and his colleagues at Duke decided to put the robot the monkey was controlling in Kyoto.

> *"We got a 5 kilogram monkey to control a 100 kilogram, 150 centimetre robot in Japan by physically generating movement out of brain waves. And we showed it could be done, it could be walking—it doesn't have to be just upper limbs anymore. And it could be done across the planet very efficiently. In other words, we scale space."*

All very remarkable, you might think. But so what? Controlling robots by brain waves is not generally something that most humans need to worry about doing either. But that, of course, is profoundly missing the point.

> *"I think that in the short run, over the next few years, the main impact of this thing we call 'brain-machine interface' will be in medical rehabilitation. No doubt about it. Patients who are paralyzed will benefit from this possibility of bypassing the lesion and using brain activity to control prosthetic devices of a huge variety: single limb, lower limb, whole body, upper limb. There is now a huge diversity of potential devices. Then there's communication. For people who cannot communicate, they will be able to use their brain activity to communicate. It's not only paralysis.*
>
> *"We're working here in the lab on prosthetic devices for Parkinson's disease that take advantage of the basic science that we've discussed regarding this new model of the brain. They would never work if the brain operated on the classic model."*

For those of us anxious to focus on the boundless medical possibilities, it's easy to gloss over that last point. But it's hugely significant as well, because Miguel is convinced that his research not only directly paves the way for a significantly enhanced quality of life for vast numbers of people, it also provides emphatic evidence that our core understanding of how the brain works needs to be dramatically overhauled.

He doesn't deny, of course, that there is a high degree of specialization of brain processing that occurs in specific regions. But his main point is that just looking at our brains as localized regions of specific activity is missing the big picture, which is all about the broad connecting circuits that are distributed throughout the entire brain.

"There is a degree of specialization, no doubt about it. But it is not as strict as we were led to believe, and is not phrenology, not by a long shot.

"This is still a tough debate, particularly in the vision community where there's long been a natural focus on a localized approach. But the evidence is growing so much. Here's another point to mention: we can find visually-driven neurons in the tactile cortex, in the motor cortex. So, it's not just in one place. And I think many people would agree with me on that right now."

Since brain-machine interfaces naturally involve combining a wide variety of different sensory inputs with other areas of brain processing, it seems the perfect arena to emphatically illustrate Miguel's central thesis of our brains fundamentally operate, a perspective which unsurprisingly has the broadest possible relevance imaginable.

"I truly believe that we're going to see other ways of brain-machine interfaces, not only for patients with neurological disorders or psychiatric disorders. We'll see this technology advance using non-invasive methods: so, no sensors inside the brain but from the outside. This will allow us to have a completely different experience of interacting with our computers and anything that is digitally controlled.

"In a couple decades we will very likely be a part of our desktops, our laptops, our iPads or whatever: we will be a part of the operating system in the sense that we will be interacting by thinking, with icons, with applications and so forth. And we'll get feedback. We are going to assimilate the operating system."

It sounds a bit out there, I have to admit. But underestimating Miguel Nicolelis strikes me as decidedly unwise: he's definitely earned the benefit of the doubt.

The Conversation

I. Revolutionary Rumblings

Beyond 1:1 maps

HB: Let me start with rats, because that seems to be where things started with your research, as you describe it in your book, *Beyond Boundaries*. So tell me about that, tell me about your first work that started with John Chapin, right?

MN: Yes. At Hahnemann University. Well actually, very few people know—and that's one of the reasons I mentioned it in the book—that this whole field of brain-machine interfaces actually started with these experiments in rodents.

John was in Philadelphia looking for a postdoc to help him and I was graduating in Brazil at the University of Sao Paulo. And we both had the same idea independently of what we wanted to do in the future. He put an ad in *Science* that was not supposed to be answered, because it was one of these ads—

HB:—where he had someone in mind already?

MN: Yes, because he had someone in mind already. But he had to put up the ad to justify hiring this person. And I answered it.

HB: What a bummer for the other guy.

MN: Yes, exactly! He got hired later anyway, as it happens. So I answered it and John just called me straight away. It was the late 1980s, and back then it was very rare to get an international phone call in Brazil, particularly at the university. Anyway I got this phone call and I could barely speak any English, but John was all excited on the phone saying that I should come to Philadelphia for an interview, because what I wrote to him was exactly what he was thinking about.

It was this idea of recording from large numbers of neurons, many neurons simultaneously. Because until then the dogma was that we would just record one at a time, out of billions of neurons in the brain, and you would find the answers.

And each of us—for different reasons, that's the very peculiar aspect of it—came to the same conclusion from very different directions. We wanted to look at what a population of neurons does when it fires together.

So John had started pursuing this, both as a graduate student and as a postdoc. He is an electronics genius, and he was building his own system. I was doing something different, but I really wanted to devote my entire career to that. I knew that's what I'd wanted to do since I was a graduate student—a medical student, actually. So we got together and we started developing this technique to record from multiple brain cells simultaneously.

And the rat model was perfect. In particular, the whisker model, because there was a lot of physiological work done on anaesthetized animals, and there was this suggestion that you had these maps — these precise maps of the facial whiskers of the rat at different levels of the brain—and that in these maps each whisker was represented very clearly by a cluster of cells.

HB: There's a one-to-one map from the regions to the actual neurons.

MN: Exactly, a one-to-one mapping. An isomorphic map, we say: the distribution of the rows and columns of whiskers in the face, into a three dimensional map at each level of the trigeminal pathway, the pathway that conveys tactile information from the face all the way to the cortex.

HB: Just to back up for a moment: I've only very recently become intimately familiar with sensory capabilities of rats. They use these whiskers to detect all sorts of things, right? They use them as a sensory device to get a sense of where to go and whether an opening is large enough or an opening is small enough—

MN: Like your fingertips.

HB: And you mentioned something before about being anaesthetized. And one of the interesting things I learned from your book is that previously when people were trying to study neural behaviour, neural pathways and signals, the actual subject was anaesthetized. There wasn't this idea of having a live subject that was there in front of you that could be probed.

MN: Yes. Most of what we knew in the beginning, in the 50s and 60s, was done in anaesthetized preparations: that's how we studied sensory systems like the visual system, the tactile system, the auditory system.

In the mid 60s a researcher at NIH, Edward Evarts, created a preparation to record from individual neurons in awake, behaving monkeys. And that was a big revolution, but it was still one neuron at a time. And people do it to this day, the preparation has survived for six decades: it's a very useful preparation.

But we wanted to know this: since you have these neurons at different levels that represent the same whisker, how do they fire? How do they respond to a tactile, mechanical stimulation of the whiskers in an awake, behaving rat?

HB: So you're able to put electrodes in the rat's brain, and you mentioned how you were inspired by Joseph Silk's book about the Big Bang to measure the neural electromagnetic signals in a distributed way by putting the electrodes all around its brain.

MN: Yes, I've long been fascinated by astronomy. And during medical school I read his book on how they linked the radio telescopes in England to measure the radio signals.

HB: So this is a form of interferometry.

MN: Yes. Of course we couldn't mirror the entire system, but the logic was very similar. So I wanted to see how different sources —neurons—interact, how they work together to underlie some perceptual capability, in this case tactile. And rats love to use their whiskers to judge openings in a wall to see if they can run through it, which is why I like to compare it to fingertips.

HB: So this is what you suggested to John Chapin: this idea of this distributed measuring device, putting these electrodes all over the rat's brain while the rat is awake to be able to measure the signals in real time, or close to real time. This was the idea, right?

MN: Yes it was this. And the key aspect there was to record from multiple structures simultaneously, which remains one of the key features of our laboratory to this day. Twenty-five years later, we introduced this approach of looking at multiple structures simultaneously and lots of neurons in each structure at the same time.

The first night we started we recorded 26 neurons simultaneously. For us, that was out of the universe, because people were recording 1. And 25 years later we're recording 2000 neurons simultaneously. I have a plot of how this progressed over time: it's really exploding now because of technological advances.

But the idea was to test a theory that was out there. It seemed very logical, very clear, that each neuron in each cluster of neurons that represented a particular whisker responds only to that particular whisker. That's the reason I suggested to John that we put these electrodes all over and he thought it was a great idea to validate a technique, to support a result that was pretty much accepted and expected in anaesthetized preparations.

But the moment we did that, even before we got these animals to do complex things with their whiskers, just having them be awake while touching individual whiskers, showed us that the story was much more complicated. In fact each neuron in every one of these clusters responded to many whiskers, not just the principal whisker that it was supposed to represent.

HB: So what happened? What did you find?

MN: Well, one of the things we found was that everything was much more dynamic. Even the clusters that were supposed to represent just one whisker, even the lower levels of the pathway, respond to many other whiskers. And they're influenced by a lot of factors.

It took us ten years. I had to move here to Duke and develop a completely different behaviour task to get these animals to use their whiskers more normally, to see the context of having the animal actually *do* something and change the way the neurons responded to the stimulation that was provided.

In fact, we published a paper showing that the neurons are firing well before the whiskers touch anything. It's almost like the brain is preparing itself to detect, it has an expectation of what is about to happen.

And this was not accepted at all in the late 80s and early 90s because the system was believed to behave just as a response mechanism, a feed-forward mechanism: you touch something and then there's a volley of activity that covers the whole brain.

And what we are showing now is that there is a huge wave of activity coming from the top-down, preparing the brain for what is about to happen. So the brain is always expecting something in the future.

When we started recording, let's say, 50 neurons simultaneously, we starting fitting in real time, or quasi-real time, the activity of these 50 neurons to better recognition algorithms, computation algorithms, to see if we could infer by the pattern of firing the size of a diameter of an adjustable hole that we created for the rats to touch with their whiskers. We wanted to see if we could get a sense, just by looking at the neurons, whether the rat would sense that it could pass through the hole or not.

And we discovered in the mid-1990s that we actually could predict, trial by trial, what was this diameter together with the animal's behaviour, with only about 50 neurons. Which didn't make any sense because, why should there be millions of cells there, if with only 50—

HB: That's a lot of redundancy.

MN: A lot of redundancy, exactly. It looked like what was going on was that we had something like a voting mechanism, like the neurons are voting somehow.

By then I had left Hahnemann and I was here at Duke already for about five years or so. John and I got together and said, "*We need a new preparation to test this idea.*"

HB: The idea that decisions depend on some ensemble of neurons all over the place?

MN: Yes, the population of neurons is voting at each moment in time and the real outcome depends on this voting. It doesn't depend on a specific class of cells or a cluster of neurons. It depends on this distributed representation.

HB: Which is very different than this idea of a one-to-one map. It's completely different.

MN: Yes: it's 180 degrees apart.

Questions for Discussion:

1. What do you think Howard and Miguel mean, exactly, by "a lot of redundancy"? How might this concern be related to broader notions of efficiency and evolution?

2. What is interferometry and how is it related to Miguel's ideas about the brain that he was keen to probe? To what extent does his anecdote about being inspired by Joseph Silk's book about the Big Bang serve as an example of how pioneering ideas in one area of science might be relevant to another?

3. In what ways does Miguel's medical background give him a different perspective from other neuroscientists who have entered the field from other directions, such as computer science or biology?

II. Plowing Ahead

Pivotal results and scientific scepticism

HB: I wanted to ask you how difficult it was actually to probe the rats. Because it sounds very easy: hit a whisker here, hit a whisker there. But presumably that's actually...

MN: Very tough. We had to learn all sorts of tricks to do that. My entire postdoc I was doing that. I didn't wear glasses before and by the end of four years I needed glasses because my eyes had gone bad just by doing this every day for hours and hours.

HB: Did you have favourite rats by the way?

MN: Oh, well, you always have. I had a group of rats named after German opera characters and I had a group of rats named after Italian opera characters.

HB: Could you compare? Was there a competition?

MN: No, no, it was just for fun. But the German group behaved much better.

HB: Well, they're more rigorous, presumably.

MN: Exactly. So for my Italian friends that was a big disappointment. But these animals were actually some of the first ones to be implanted with sensors in multiple locations of the brain. When we published this paper in 1995, it was a shock: nobody was expecting that.

It was published in *Science* and today has 500 citations. But at the time it met considerable resistance. Part of this was because we were using a sensory system, a tactile system, and your animals cannot report what they're feeling.

So people said *"OK, you've found this neurophysiological evidence, but what does that mean in terms of the animal's experience? How can you predict the animal's behaviour if you can't get a real quantitative measurement—besides number of trials correct—of what is really going on?"*

Basically, the problem is that you cannot interrogate your rats.

HB: OK, but that's a bit confusing to me. Because the story that I remember you telling in your book was that you managed to find very strong evidence for this distributive model, which went against the prevailing dogma that most adhered to.

So your results weren't terribly well received at first. There was a lot of scepticism, which led you, in fact, to go to the next level which we're going to talk about in a moment. But I remember thinking as I was reading this, *"Well, why weren't they very well received?"*

OK, there are academics who are holed up in their own world of past beliefs, just like everyone else—we've all been through that, we all understand that. But it seems like these were pretty compelling experiments that you did to support your claims.

So if I'm a one-to-one map, localized, one-whisker-one-neuron person and then somebody comes and shows me some fairly significant evidence that argues for a more distributive-type model, what can I say to that?

I mean, maybe I'm missing something, but the idea that we don't know what the rat is feeling doesn't seem like it's terribly relevant to the main issue at stake here.

MN: Yes, well, we heard all sorts of things. Because you have to realize that this is not physics. The physics community is much more open to breaking dogmas, I think, from my little experience with physicists.

HB: Well, I would take issue with that. But that's another story.

MN: For biology, there was a lot of vested interest in this model. In a general sense, this model was awarded a couple of Nobel Prizes. It was difficult, very difficult, to be saying something against it.

HB: So were people giving you any clear feedback in terms of, *"You didn't do the experiment correctly"*?

MN: No, not so much that. The papers were being accepted in big, reputable journals: *Nature, Science, Journal of Neuroscience, Neuron*, all the big journals in the field. It was just that people were not accepting it. They were very afraid of a new technology, a new way to look at the data. We were doing statistics for the first time in neuronal responses because we could get hundreds of trials.

You have to realize that when we published our first big paper in 1994, people were doing ten trials, fifty trials per whisker because you couldn't hold the cells for too long: you had to hurry. But since we could hold the cells for days, we were doing five hundred trials. So we had to get the statistics on 360 trials, I mean I remember that very vividly. One of our reviewers asked, *"Why so many trials?"* And I replied, *"Well, we need to do statistics."*

HB: And why not? That, too, is an odd sort of response to me. I mean the more you have—

MN: Of course. But what I'm telling you is that there was not a great tradition of quantification and statistics in the field. A few theoreticians loved it because that was what they had predicted. But many experimentalists had a lot of difficulty in accepting it.

HB: So there was this sense of, *"Here come Nicolelis and Chapin with their fancy mathematical models that we can't really understand"* or something like that.

MN: Yes. And it was done in rats. At that time there was considerable hubris from the people who worked with primates. They thought,

"Just because you find these things in rats, that doesn't necessarily mean that they apply to what we're doing—"

HB: Oh really? There's a strict hierarchy?

MN: There certainly was. That was a big deal when I was a postdoc. I had several, very famous, monkey physiologists saying to me at meetings or in my talks: *"Oh, this is very nice but these are just rodents. You have no evidence for any of this in monkeys."*

HB: So faced with this response, you decided...

MN: Well, John and I joked that we decided to move just a couple of millimetres ahead to the motor cortex. That's another thing you have to realize: until very recently people who worked in the motor cortex didn't talk to people who worked in the tactile cortex. They're just a few millimetres apart in the monkey brain, but it was divided into two very different, sociologically separate, groups.

That was already almost something of a taboo, for tactile people to cross the central sulcus and start recording right in front of it. But John and I always liked those adventures, so we went ahead and started doing exactly that. And that's when the brain-machine paradigm that you hear now, that lots of laboratories are using now, was born.

There had been some experiments in the 60s done by a phenomenal scientist, Eb Fetz, that were somewhat similar: he was trying to look at how he could condition individual neurons with feedback.

But what we proposed was very different. We wanted a quantitative way to demonstrate that distributive coding is the way to represent motor information to generate motor behaviours.

HB: So what did you do? And what happened?

MN: You have to realize that this was 1996. At that time there were only a handful—probably 3—of labs in the world that could do simultaneous multi-electrode recordings and only our lab was doing them at multiple sites.

So we said *"OK, instead of just collecting this data and analyzing it off-line, let's use this little experiment that we just finished with the rats as our new paradigm. Let's do a real-time analysis and test the main predictions of the distributive coding theory by seeing whether the animal can use its brain activity alone to make an artificial acti-vator reproduce the kind of movement needed to perform a task and get a reward."*

At that point nobody was talking about that. I was here at Duke and was starting to record for monkeys. I had just published a paper about multiple sites in monkeys. And we agreed that we should first do the rat study to see if it was viable, because of course you can test ideas in rats. But at the same time that we were seeing that the rat work was moving ahead, I started with the monkeys here. So we published one paper after another. We published the rat paper in 1999, and seven months later in 2000 we published the *Nature* paper.

But the funny part is that the first paper on rats we sent to the big journals—*Nature, Science*—they all rejected it without any justi-fication. So we sent it to a new journal at the time called *Nature and Neuroscience*, which is now very prestigious, but at the time was just starting out.

Over the years I've done that many times: I have these papers that the big journals look at and say *"Oh, we don't want it."* So I send it to a start-up and all of a sudden they have lots of citations ten years later.

When that *Nature and Neuroscience* paper came out the results came as a big shock to many. These rats were using their brain activity alone to move a one-dimensional lever: a little robotic arm to bring water to them so they could drink.

Here's how it works: the rat is implanted, and he's trained to press a bar to get water. So if the bar is pressed, a little arm—a one-dimensional arm—collects water from a dripping source, and brings the water to its mouth.

Now at this time we're recording 46 neurons in the motor cortex of this rat, and we use a very simple algorithm to transform these electrical signals into an output that would make this one-dimensional movement be generated by the robot without the need of the animal to actually press the bar.

And the most astounding finding of that study—for everyone, including us—was that during certain periods the animal would *get it*. He'd stop moving his forepaw to press the bar himself and just *think* about it, getting the arm to go there, stop, collect the water and bring it back to him so he could drink.

And for periods of time, a few minutes, the animal would be able to do that successfully. And then he would go back and try to press the bar again himself, but at that point the lever was disconnected so nothing happened.

So when we published, some people said, "*This is funny. It's just a few seconds, a few minutes sometimes, that the animal stops moving. Maybe it's not working in the way that you suggest.*" We suggested that the animal was somehow learning, that he could rely on—

HB: Just on his thinking.

MN: Well, he didn't know that it was his thinking. But he was learning that he could rely on something that was non-motor behaviour. Because the muscles did go quiet.

Anyway, I was doing the monkey work in parallel, and we saw that the monkeys stopped moving too. We saw that first with owl monkeys, a small species of New World monkey that I was very familiar with because of my previous work in Brazil. And then we heard people say, "*Well this is not a monkey, this is a New World monkey*". Of course it's a monkey. It's a primate. But this just shows how complicated this whole business is.

So we moved to bigger primates—Rhesus monkeys—all in parallel. And when we got to the Rhesus monkey—that was in 2003—we were using seven degrees of freedom of reaching and grasping: one dimension of grasping and six degrees of reach.

And we have movies as part of a complete documentation of what was going on. The animal would get it and stop moving to play the game they were supposed to play, using brain activity alone to move this arm to control a little computer cursor that gets to targets. And that was when everything exploded.

HB: Now you're dealing with primates, you're dealing with a higher level of intelligence: you're dealing with beasts that can be trained.

MN: Yes. They cannot report verbally to you, but what they can report behaviourally is much more elaborate than what a rat can do. But at that point we start seeing the same old thing. It was sent to *Science*, two reviewers accepted it and one reviewer didn't. *Science* declined to publish it so we sent it to *PLOS Biology*, which was then in its first issues, and to this day this paper is the 11th most cited paper in *PLOS Biology*. That is very important for us. It shows that it's not **where** you publish, it's **what** you publish— something I've been telling my students for 30 years now.

HB: Eventually it comes out.

MN: That's right: eventually it comes out. But it also shows that people can really triage what is good and what is bad. It's not very straightforward.

But anyway, back in 2003 we were recording 100 neurons. So very quickly we went from 12 to 26 to 46 and then 100. We established the paradigm. And in 2004, only one year after, we had the first multi-electrode recording paper, with 11 patients intra-operatively showing that the same algorithms that we had used a year before with the monkeys would work with humans.

We could get patients in the operating room who were conscious: they were awake because after the neurosurgeon penetrates the brain there is no pain anymore. The neurosurgeon required the patient to help in certain manoeuvres there, to locate certain parts of the brain. So the surgeon is always asking what the patient can report.

We had 10 minutes during that surgery to introduce a new electrode that was actually very beneficial for the surgery itself, to record from about 50 neurons in the deep structures of these 11 patients, and actually see the real time output of a one dimensional game that they were playing during the surgery. And we found the same thing.

We reproduced that recently with 25 more patients last year, and now we have a library of 46 patients recorded where we can see the same physiological findings that we found in monkeys.

But the funny part is that in the beginning nobody believed that this could happen. We never dreamed that this could become a tool for medical rehabilitation. We were just doing basic science.

Questions for Discussion:

1. Are you surprised to learn about a hierarchy between researchers on different animals? To what extent is it reasonable to be suspicious that an experimental result for rodents would somehow apply to primates?

2. Are scientists more sceptical than other professionals of approaches that go against "conventional wisdom"? Are they particularly trained to be that way? If so, in what ways is this a good thing?

3. Does the scientific peer-review process work as well as it could or should? (Those interested in a more detailed historical perspective on scientific peer-review are referred to Chapter 6 of the Ideas Roadshow conversation, **Science and Pseudoscience***, with Princeton University historian of science Michael Gordin)*

III. Interface

Powering robots with monkey minds

HB: I want to get to some of the applications, but before I move on from the theoretical picture, I have a few more questions.

First a summary: there are these electrodes that are implanted in the brain in a distributed way, whether it's a primate or a rat or humans or what have you. They're recording all sorts of signals and I need to have all sorts of processing power and very clever algorithms to be able to interpret those signals in a particular way.

The first set of experiments with the rats gave us a strong argument for this notion of how information processing in the brain might be distributed. And we discussed some ways we can capture the signals and get some clear evidence that animals, whether rats or primates or humans, are able to physically affect some sort of device using their brains.

MN: Yes Let me add one thing before I forget. When we started, people thought that, given the non-linearities that we have in the brain, the computational algorithm that we would have to use in real time would be so complex that it would simply never work. So another big surprise is that we did all of this in the beginning—and we're still using it, actually—with a linear algorithm. It was a multi-linear regression analysis, a Wiener filter. It's just a multilinear regression. And it worked.

And it worked because we gave some percent of correct predictions for the brain, and the brain adapted itself to the limitations of the algorithm. It's almost like the brain was our supercomputer dealing with the limitations of our algorithm. For ten years people have put a lot of emphasis on trying to find the silver bullet algorithm to do a brain-machine interface. And it turns out that, up to a point, they're secondary.

And they're secondary because you are appealing to what the brain can do for you in terms of plasticity. The brain, I like to use this metaphor, is the only orchestra that changes its instruments as it produces the desired music. So the very music that the orchestra produces is actually shaping up new properties of these instruments.

So by taking advantage of this in a very simple way, we didn't need to use time consuming models or algorithms. We could use something that was computationally very efficient, optimal, and could fit the amount of data we are collecting. We can do that in the same window of time—a third of a second—that the brain usually uses to plan a motor behaviour in the future.

HB: So it seems like the brain is compensating on the nonlinearity side somehow, and then you can use a linear algorithm because the brain's taking care of that—

MN: Exactly. Basically, if you use random algorithms, random regressions, it doesn't work. So you need to give some variance explained by the algorithm—30-40 percent—and the brain does the rest of the job so the thing would work. And it does that by changing its own physiological properties. That's what we documented.

HB: I want to get to the plasticity shortly.

MN: Sure. But getting back to what you were saying before, what happened was that after this first wave of papers everyone got into the "upper limb brain-machine interface" bus. To this day there are people who continue to do the same paradigm, without any change. But we decided that that's fine, but here in the lab we have this tradition of trying to push the envelope, trying to do things that nobody would dare to because they're too risky or too complicated. So we wanted to see if we could generalize these results to the lower limbs. And actually to this day we're the only lab that did it.

So we started out with an animal walking bipedally, like we walk. Rhesus monkeys walk on all fours, but if you suspend the chest, if you give them support, they will walk bipedally—on a treadmill for instance. And we saw that we could do the same thing: we could record from their brain activity and predict the step cycle of these monkeys even better, for the same number of neurons, than we could do for upper limbs.

Shortly after that, I was in Japan and met a phenomenal Australian-Chinese roboticist named Gordon Cheng. He had just moved to Japan and was working with an American company to build a state of the art humanoid robot in Kyoto at ATR laboratories.

But he lacked one thing: Gordon wanted to get his robot to walk autonomously, but he didn't have the brain, of course. So we decided to do a crazy experiment in which we lent CB1, the name of his robot, a brain—a monkey brain.

We got these monkeys to walk here at Duke, and Gordon set up a really high speed internet connection that would allow the brain activity that we recorded in real time as the monkey walked to be sent to Kyoto, so that CB1 could walk, guided by the monkey's motor commands. And at the same time, video footage of the walking robot could be sent back to Duke so that the monkey could see it.

HB: The monkey gets a feedback, a visual feedback.

MN: That's right. He has a huge screen in front of him as he's walking and he sees these legs walking. Every time the robot touches the ground the monkey gets a reward, a piece of fruit or a little bit of juice. So the monkey got conditioned to imagine this walking even *after* we turned off the treadmill. So even after the monkey stopped moving its own body—

HB: He was thinking about it.

MN: He was thinking about how to get those legs moving in Kyoto: we could get the robot walking for an hour without the monkey moving here at Duke, just by rewarding the monkey thinking about it.

We also got that trip around the world faster than it takes for the action potential to leave the monkey's head and reach the muscle of the monkey. It was just 20 milliseconds faster, but it was faster.

HB: That's pretty impressive.

MN: Yes. And of course we got a 5 kilogram monkey to control a 100 kilogram, 150 centimetre robot. In Japan.

HB: But *just by thinking*.

For me, ***that's*** clearly the most remarkable aspect of all of this. It's one thing to say, *"OK, you're a scientist and you're in a lab and I can see that you're controlling a robot over here by some form of remote signalling. Yes, I can see you have an impressively super-fast internet connection, too, and that's wonderful,"* but, for me the mind-boggling thing is how that signal is actually being generated to make all this happen—the idea that ***after*** you turn the treadmill off, the monkey's been conditioned: he realizes, *"Gosh, if I keep thinking about moving my legs, this thing on the screen in front of me keeps moving and I keep getting my juice."* So he *keeps thinking* about moving his legs after the treadmill stops and the robot *keeps moving*. In Japan.

MN: It is physically generating movement out of brain waves. And we showed that it could be done by walking—it doesn't have to be just upper limbs anymore. And it could be done across the planet, very efficiently. In other words, we scale space. So from that point on, I like to say that the actuator does not need to be next to the subject that is controlling it by brain activity.

HB: That's showing off, though, to put it in Japan. I mean, you could have just put it around the corner.

MN: Well, yes. But we've done this since the first monkey paper. In our first monkey paper in 2000 we had a robot arm at MIT at the same time that we had one here at Duke. We got both robots synchronized, so that we could show that even with the delay of the internet that we had to compensate for, the robots could perform the same task no matter where they were.

Neuroscientists didn't appreciate that very much but the robotics community liked it very much, because we're showing the potential of teleoperation by brain control.

HB: I've mentioned this before in other filmed conversations: I'm pretty sceptical of interdisciplinarity as something imposed from without. Sometimes you'll have a university administrator who says *"And now we're all going to be interdisciplinary! We're going to put a historian, and a neuroscientist, and an economist together, and wonderful things will necessarily happen".* And very often, of course, nothing happens. They may have a very nice lunch or something, but that's about it.

But it seems to me that there's a natural type of interdisciplinarity to what you're doing.

MN: We joke in the lab that this is grassroots multidisciplinary research. Because we have a need for a roboticist, we go find one. We often need engineers, mechanical and electrical engineers, so we have them too. Our lab is formed from a lot of backgrounds: computer science, neuroscience, psychologists, physicists. We have a pretty multidisciplinary approach by need. It was not imposed top-down.

I think that created a perspective for us that is very difficult to mimic and very difficult to understand outside of our lab. When I give talks and try to tell people what we're doing—even to my colleagues in other departments—I notice that there is often a real difference in cultures because people have generally been trained to specialize in small, focused questions. That can be very fascinating, but it has its limitations. I was just talking to a physicist a couple of weeks ago at Cold Spring Harbor and we both agreed that reductionism was not going to explain the brain.

Questions for Discussion:

1. Why do you think so many people might have assumed that brain-machine interfaces would only work with upper limbs as opposed to lower limbs?

2. To what extent do you think this type of research will eliminate some of the traditional boundaries between academic disciplines?

3. Might it be possible to use this type of research to develop better communication techniques with animals?

IV. Against Reductionism

The pernicious influence of physicists

HB: I kept sidetracking you when you wanted to talk about of localized versus distributed and reductionism, so let's talk a little bit more about that now. I have some questions. I have a physics background, as you know, so I have reductionism in my genes.

MN: Of course. Because it works.

HB: Exactly, because it works. Which doesn't necessarily imply that it works in all cases—which it clearly doesn't. So let's move to the brain, but before we do let me try to state the issue as clearly as possible. Here's my take on things.

The old physics way of thinking is: you take the simplest possible models, you look at things at a one-to-one level, and you just scale up to get a picture of much more complicated things that are made up of those basic building blocks. So you try to understand the details of one particular atom, say, or one particular molecule, and once you've done that you turn your attention to more complex phenomena and imagine them as being collections of these particular things that you understand.

That's oversimplifying things—there are all sorts of statistics that come into play, and there's chaos theory, and all sorts of other things. But that's the basic mindset, I think: looking at the tiniest individual particle and examining its properties and then from there we'll understand the grand theoretical framework.

And my understanding from reading *Beyond Boundaries* and talking with you is that for the longest time, that's the way a lot of biologists and neuroscientists were looking at things on a neuron level: we just look at the neuron very carefully, and this leads to this idea of everything being localized. You have one type of neuron in one part of your brain that controls one thing, you have another type that controls something else in another part. And it is this sort of thinking that gives you these one-to-one maps that we were talking about earlier like with the rat whiskers.

And you think something quite different from that.

MN: Yes. In fact I like to say that we biologists are victims of physics' success. But a lot of physicists don't realize what you just said: that this approach doesn't actually work for everything in nature: you can know everything about the atoms and molecules of my door, but you're never going to be able to tell from the elementary properties of those atoms, how that door is actually moving—that jump, from one level to another, cannot be done.

In fact, you cannot, in a very analytical way, even predict the movement of three bodies interacting, let alone billions of stars in the universe.

HB: Yes, that's embarrassing. We like to keep that under the rug, actually.

MN: Yes, but there are several things that are kept under the rug because of the success of high-energy physics. But most of the phenomena that we see around us are non-computable.

HB: That's why we promote high energy physicists you see: that's why they're the poster boys of physics.

MN: Yes, I love high energy physics, don't get me wrong—and physics in general.

When I was younger, I was hugely influenced by Carl Sagan, and I still think a lot about this. In one of his books, I think it was *Cosmos*, there was this phrase, "*In the beginning there was the universe*" or "*there was the Big Bang*" or "*there was physics.*" I'm actually trying to change that in my own way, because even physics depends on our primate brain.

Physics may simply be the best way to describe the limits of the logic that this biology can produce.

HB: I see where you're going here—you're opening up a whole different debate. We should end with this because it could go on for hours.

MN: But this is beautiful, because this is what our research is now bringing about. So I would have to start a new version of this *Cosmos* book with *"In the beginning, there was a primate brain."* Because that's the constraint of all science that we do.

HB: Because we can't get outside of our own system.

MN: We cannot get outside. There's this famous debate where Einstein and Gödel are debating the equations of general relativity. Einstein thought that it was the theory of everything. But Gödel goes to him and says *"But you are not part of those equations, you cannot be."* And it's true, his brain was not there. Those equations could not explain how the brain works. And the way the brain works generates the logic that allows us to do science and to create our best shot at explaining the universe.

And believe it or not, I think this is intimately connected to when we move to the question of a localized to distributed approach. I'll tell you why. Why is it so difficult moving from the engineering based diagram of the brain, where every part of the brain has a function?

It's very difficult because of our tradition, because of the success of determinism and physics, we cannot imagine that most of what we do emerges as emergent properties of the system.

HB: OK, so let's talk a little bit about emergent properties, because here are the obvious questions, I think, that a standard reductionist might throw at you.

I'm guessing that this reductionist would say something like, *"OK, fine: we understand that there are different levels of phenomena, different things going on. And some things can be successfully described with one language and other things can be described with other languages. But then you have the obligation to tell me when I pass from one domain to another. You have to describe, to some extent anyway, where that threshold is.*

In your book you talk about how you shouldn't think of the neuron as the fundamental thinking unit, but instead look at populations of neurons.

MN: Yes.

HB: So my questions to you would be, *Well, what defines a population? How big does it have to be? When do you know that you've reached this critical point? And what's going on, exactly, that accounts for this special emergence property that comes out?*

MN: Those are the questions that actually define our research program. When I give a talk at any university, that's basically the first slide.

There was this Canadian psychologist I've read a lot about over the years, Donald Hebb, who suggested in 1949—way before we could actually do the experiments—that we should be looking at populations. And to me, the key question is, *How many neurons?*

We know that some invertebrate brains can work with a few hundred neurons to still produce behaviour that is pretty well adapted. But we are far from having a comprehensive understanding of how a C. elegans' brain works, although many are now realizing that this is also, probably, a distributed system. We cannot predict how neurons are going to fire in very simple behaviour, let alone talking about a monkey moving a robot arm.

So that's what motivates us, exactly the questions you describe. The hypotheses that we make derive from these very questions. How many neurons do you need? Is it always the same neurons? Perhaps if you have enough of them, they are exchangeable: it can be any neuron in that pool.

HB: So it seems to me that there are a few basic beliefs that fit together to account for your world-view: you've got this key notion of distribution; you've got neuroplasticity—it's not as if I have one neuron that is doing the same thing all the time and always has, the function actually changes—and then, my sense is that your two beliefs of distribution and plasticity somehow combine to get this brain-centred view, as the brain is evolving.

MN: Yes. And the final item is a probabilistic view of the brain, in the sense that since you have billions of elements, I see the neurons as probabilistic elements. They can be exchangeable.

And from this probabilistic cloud of activity you come and generate the deterministic behaviour, which is something difficult for biologists to grasp: that you can ride a bike so precisely, or make a jump shot, or score a soccer goal when the underlying mechanisms that are generating that very precise deterministic mode of behaviour are all probabilistic.

HB: And that's because of the numbers presumably, because there's so much complexity that's actually going on.

MN: Yes.

HB: But I'm guessing it's not *inherently* probabilistic—this is often my problem in general when people invoke probabilistic arguments (and another reason why I tend to stay far away from discussions of the foundations of quantum theory, but that's a whole different topic).

For example, if I want to measure the pressure of a gas or something, that's probabilistic too: I've got molecules that are zooming all over the place, some of them are going here, some of them are going there.

In principle, if I were God, I could tag all of these things and get a clear sense of everything, but the probabilistic arguments that I'm using are forced on me because I can't *in practice* do any of that stuff, so I have to settle for probabilities.

MN: Yes. That's a good analogy actually.

HB: But what's actually happening here in this case? Is it like that? Or is it somehow a little bit different?

MN: Well, as I say, it's like a voting system. So you need to elect a president of the brain. And you have one round to decide it, you have one shot, or otherwise you may die. So this system has to be optimal, resilient to lesions, so if you lose a few elements everything still has to work.

So to briefly summarize more than 25 years of work of many labs, not just mine, what seems to happen is that the brain has to achieve this goal. And to achieve this goal, it needs a certain mass of neurons to cooperate in a given moment in time to generate the motor program that will give rise to the behaviour.

It can achieve that by a huge number of different particular combinations of neurons acting. In fact this number of combinations is so gigantic that it may never repeat itself during the entire life of a subject. So my prediction is, if I move my arm like this, for the rest of my days, the kinematics are the same, the movement are identical at the limit of my muscles. But the specific neuronal combinations that I might record that would underlie this behaviour would never repeat themselves.

HB: In other words, there's an aggregate signal which is being produced somehow which is the same, but on the micro-level the details of how it's being produced is going to be different every time.

MN: Yes, exactly. And that's the reason why a single neuron can not be the functional unit of the brain. Because it's just a peon in the game of producing this behaviour.

HB: It's just a mode, a mechanism, for producing the signal. It's the signal that's everything, and how it gets produced doesn't really matter as much.

MN: The brain has to achieve the goal. Because the goal is relevant to keeping that subject alive, it enhances its chances to survive. And that has to be maintained at any cost. In fact that might well be the reason why you have so much redundancy. Because you may lose huge chunks of your brain, and still be able to realize the behaviour.

Questions for Discussion:

*1. Can you give other instances where standard reductionistic approaches fail to provide a comprehensive scientific understanding? (Those with a particular interest in this topic are referred to two Ideas Roadshow conversations—**Pushing the Boundaries** with Freeman Dyson and **The Problems of Physics, Revisited** with Tony Leggett—to experience two renowned physicists exhibiting strongly anti-reductionistic tendencies)*

2. How do you think that a "die-hard reductionistic" would respond to Miguel's views expressed in this chapter?

V. Incarnating Our Surroundings

How the brain blurs "outside" and "inside"

HB: I'd like to follow up with something else that you mentioned in *Beyond Boundaries*. When a monkey is doing some particular activity, say, you can monitor the behaviour of a group of neurons, and say *"OK, these neurons are firing"*.

But then when the monkey is not actually physically doing that activity, but just *thinking about* doing that activity, completely different groups of neurons are now firing that were not actually firing before.

My sense is that this shows two things.

First of all, as we've talked about at length, once again there's this argument for distribution: different patterns of neurons are doing things.

But I'm also thinking that it demonstrates a form of neuroplasticity by illustrating how the brain might be evolving by expanding and adapting new tools and new techniques. In other words, these neurons that weren't firing before somehow become part of the "tool system apparatus" as the brain evolves. Is that a fair way of putting it?

MN: Yes. In fact our latest hypothesis that we are testing avidly in the labs here, is that the sense of self, that is pretty unique to us and probably higher primates, this belief that we are unique, individual, doesn't end at the limit of the physical body that we inhabit, but includes every single tool that we normally utilize in our daily life.

We have very good evidence for that, both from our lab and others, because as we start using tools and becoming proficient—even a monkey—cells in your brain start responding to that tool. And you start representing the properties of that tool inside your brain.

But we go a little further—and that's what we're testing right now—that this sense of self also includes the social acquaintances that we make throughout life.

So the people who are close to us, are, as the song lyrics of that Cole Porter song go, "*I got you under my skin*".

I believe that this is literally true: that there are representations of our social acquaintances that have basically merged with our sense of self—they're part of our sense of self and our identity is associated with these social relationships. And that's the reason we are so socially dependent. That would explain why we are so avid at maintaining social contact.

HB: But this is a general statement for tools of all sorts, it seems to me. Right now you're talking about how this process might provide a brain-biological picture of understanding our social tendencies, but in *Beyond Boundaries* you also talk about the brain's remarkable ability to harness physical tools. There's this image of a Roger Federer wielding his tennis racket as an extension of himself. We use these words metaphorically all the time. We say, "*It's just as if the racket is part of his arm*". But to some extent it's not actually a metaphor at all—that **is** actually what's happening on a neurological level.

MN: There are experiments showing that if people play tennis for an hour—not very poor players but those who are proficient—and after an hour you ask them, blindfolded, to point with their non-racket hand to the mid-point of their racket arm, they will take the racket into account.

And then you get all sorts of subjective reports on how very high-performance athletes feel. The good soccer players don't feel the ball per se, they feel as if the foot and the ball have become a sort of hybrid.

HB: And the claim is that, neurologically-speaking, that really **is** what's going on to some extent, right?

MN: We have a paper where we clearly show that using avatars. When monkeys play with avatars of themselves, and the avatar is doing something, we now have evidence that the avatar arm, which is a virtual arm—it's a piece of code—is becoming embedded in the body representation that the brain carries of that subject. So the animal actually acquires cells that are responding to the avatar arm, in addition to the regular arm at the end.

Here's another example: smartphones. In a relatively short period of time they've become so much a part of our culture that people now have withdrawal syndrome if they're not using their phone. If an alien would come to Earth, it would report back that, *"The most common behaviour of humans is to look at their hands."*

But that can only happen if your brain is adapting so quickly, that for any technological advance, parts of your brain are mimicking some of the steps that the machines are creating.

HB: So it's become a part of you to some non-trivial extent.

MN: Yes, there are all sorts of studies about how kids' brains behave these days because of the internet, clearly indicating that the internet is shaping the way our brains operate.

HB: In fact, the way we do *everything* shapes it. That's really the point.

MN: Yes. The introduction of books, radio, TV, and then the internet, these are going to be known in a few centuries as key landmarks in shaping human brain evolution.

Questions for Discussion:

1. To what extent should public policy decisions on education and youth development be informed by a deeper understanding of neurobiology?

2. How might the contents of this chapter be used to argue that "behavioural" addictions such as gambling are every bit as "physical" as cigarette addiction?

3. Might it one day be possible to deliberately harness tools to make our brains less resistant to harnessing tools?

VI. Imposing Representations

How the brain interacts with the world

HB: I'd like to talk a bit now about phantom pain, because it seems that this is also related to how the brain somehow imposes itself on the world. I think most people have some basic idea of what this is: after someone has had a limb amputated he still feels pain, often quite high levels of pain, where the limb used to be.

MN: Phantom pain is one of my favourite subjects that we could talk about for hours. It's been reported in literature since medieval times. Ancient civilizations knew all about that, but in Europe, in the battlefields, it was very common because you had all of these traumatic amputations.

HB: Napoleon contributed a lot to that.

MN: Yes, an awful lot. But even before, in the 16th century, the French physician Ambroise Paré was afraid of reporting it because he thought people would think he was nuts.

HB: It's the same scientific publishing problem all over again.

MN: Exactly. So he published a book in French—instead of Latin, the usual language of science at the time—describing it. The name "phantom pain" was coined centuries later during the American Civil War. Three days after the Battle of Gettysburg, a Philadelphia physician realized that both Union and Confederate soldiers who had suffered amputations were not only reporting pain in their missing limbs but the sensation that they were climbing the hill over again in Gettysburg. They were in their beds and they had the sensation of running.

Up until the 1950s, people thought that phantom pain came from something peripheral: a lesion of the nerves, or the idea that scar tissue that builds up would somehow stimulate these nerves. Again, it was all very much in keeping with theories of the time: that our brain is decoding something external from the world.

HB: Right. Passively receiving signals from outside.

MN: Exactly. In fact, to this day, if you open up any textbook you will see something like, "*We see things, we break these things into separate features, the features then go to each part of the brain and somehow* (magically) *they come together and we see*".

There's no time for that. You couldn't survive a lion coming in the Savannah trying to eat you alive because there's no time to break anything down.

But phantom pain gave us the first hint of something that didn't match that theory. There was a series of studies in the 50s and 60s, involving people being born with no limbs and reporting phantom sensation.

HB: So there goes your "scar tissue" argument right there.

MN: Exactly. And some people didn't believe those results, so there was a big fight.

But interestingly enough, if you now change the framework, and you put the brain in the center of the picture, you find a completely different explanation for phantom limb pain. And that is that the brain has an internal model of the body that it's used to, which it has developed over the years. It has ingrained that in your memories: that you have two upper limbs, two lower limbs, ten digits, ten toes. And the brain is continually verifying that hypothesis, because you touch things, you get things, you walk, and so on. Now what happens when, suddenly, you lose part of it? Now the brain has an internal model that is mismatched by the periphery and it is this mismatch that generates the illusion that you still have a part of yourself that has disappeared.

HB: So the brain is imposing somehow that you have legs, even if you don't have legs?

MN: The brain is the true creator of everything for us. Including our myths of how we came about, where we come from...

HB: I told you: we'll get to that at the end.

MN: We'll get to that at the end. I keep provoking you.

But in the phantom limb sensation, there is something equivalent that is as scary. If you have a lesion in the posterior parietal cortex in the right hemisphere, a big lesion or even a transient lesion, the left half of your body in the universe disappears. It's called hemineglect syndrome—the person neglects the left side of the universe.

I saw this in a patient in the University of Sao Paulo when I was just 22 years old. The test for the patient was the following. He had to walk in a hall, the neurology ward, and he had to turn left to go to the men's room. We told him, *"Go, find the men's room and get in there."* So he went all the way to the end of the hall, turned back, came back and turned right to go into the men's room because left didn't exist for him. Even though he had an awareness that there was a men's room on the left part of the universe, he had to go to the end of the hall to come back and walk into there.

HB: So he was conscious of where it was but the brain somehow imposed this idea that...

MN: ...it didn't exist. Right. And he would put his pyjamas on only on the right side, we had to dress his left. We would put his left arm in front of his eyes and he would say, *"That's not my arm, it belongs to somebody else."*

HB: Wasn't there also something about an astronaut?

MN: Yes. There was an astronaut who came to give a talk here at Duke many years ago, in the 90s. He told me that in one of his missions the commander, the pilot, had developed left hemineglect syndrome, and in the middle of taking off, he announced, *"Guys we have a little problem: I can't find my left arm"* or words to that effect. Maybe it was *"Whose left arm is that on the dashboard?"*

So everybody looked at each other with considerable anxiety. You don't want to have a pilot on the Shuttle who doesn't know that his left arm is actually his.

HB: I don't want to have a pilot on a normal plane that doesn't know that.

MN: Right. But it was transient: it was probably because of the gravity, it may have been anoxia, some kind of temporary hyperperfusion of the right hemisphere. When he returned everything was fine. But all of these types of phenomena give us hints that the brain is actively creating reality and is not passively decoding anything.

When you look at the two separate historical processes: the evolutionary history that generates the plasticity and the assimilation of tools, which is inherently unpredictable—it's a pseudo-random walk—and the personal history of each one of us that creates this internal model of who we are to ourselves, it's clear to me that neither of them are computable. To think that you would be able to replicate all of that in a Turing machine, in a regular computer, is ludicrous: you simply can't.

HB: Because it's contingent on history.

MN: It's contingent on history. And neither historical process can be replicated in an algorithm, in a Turing tape, in a sequential series of steps. But these two things do create an internal history, an internal model that imposes on our senses.

So I like to say that we see before we watch. So if you see something happening behind my back and should out "Look!" even before I turn around my brain is already creating a hypothesis, because it knows this room, it knows what is behind me, and it is trying to create a hypothesis before I even take any data from the environment.

HB: So again, it's imposing on the world rather than just receiving. It's this notion of imposition.

MN: Yes. For those who like philosophy it's Kant to the limit.

Questions for Discussion:

1. When Miguel talks about the internal model our brain creates not being computable, is there a distinction between not being computable in principle or in practice?

2. How can the theory of the brain imposing its model on the world explain how people born without limbs can sometimes feel phantom pain?

VII. Distributed vs. Local

The big issue

HB: Let me be devil's advocate for a moment. You've given me all sorts of evidence to believe that information is processed in a distributed way in the brain, convincing me that there is no 1:1 map between neurons, or specific groups of neurons, and brain functionality.

But on the other hand, I keep hearing that the brain is also divided, roughly, into different areas: the motor cortex, the tactile cortex, and so on. Clearly these are different parts of the brain that are used for different things. So how can that square with your ideas about everything being distributed?

MN: Well, it's not homogeneous. It's not distributed homogeneously all over. During development, some areas of the brain receive input from the periphery of some sensory organs. For instance, in one part you receive mainly visual input; in another, auditory and tactile. Of course, this is how the brain is put together: there's a scaffolding of axons that shape up the functional distribution of the signals.

Yet, we now know that if I blindfold you—a normal subject, no visual problems—in a few minutes all the tactile responses are in the visual cortex. Nothing has been created *de novo*, since I just put your blindfold on: there's no time for genes to be expressed, proteins to be reprocessed.

We have evidence in animals that even in the primary fields that we all know—visual, auditory, tactile—even these fields can respond to other sensory models. So there is a specialization on top of a broad background of multi sensory processing going on there. So that's my response there.

The other thing I was expecting you to ask me was, "*Well, if we lesion some specific part of the brain people go aphasic, isn't that an indication of things being localized?*"That's the classic deniability hypothesis for those who advocate localization, right?

HB: Right. I was getting to that.

MN: Broca, the famous French physician noted this effect with one of his patients at the end of the 19th century. But we now know that for that to happen to such an extent, you have to destroy the underlying axons, the fibres that are crossing everywhere, connecting many parts of the brain.

HB: So you're isolating it by cutting out the connections.

MN: You're cutting out the circuit. So if you have lesions of the grey area only—only the neurons—you can lesion a lot, a huge chunk, and you are not going to see symptoms. That's very common in superficial strokes.

HB: Because the connections are still there.

MN: The connections are still there. So you need a grid for everything to work properly. It's pretty much like the internet. If you take a few peripheral servers out, nothing happens. But if you were to take *Google* out, all the servers that Google has, the internet will feel the effect. Because now you have a massive disconnection of a key hub.

So in my view none of these things is incompatible with the distributed hypothesis. Classically, these examples have been used to suggest a localization of function: there is a degree of specialization, no doubt about it. But it is not as strict as we were led to believe, and is not phrenology, not by a long shot.

So this is still a tough debate, particularly in the vision community where there's long been a natural focus on a localized approach. But the evidence is growing so much. Here's another point to mention: we can find visually-driven neurons in the tactile cortex, in the motor cortex. So, it's not just in one place. And I think many people would agree with me on that right now.

Questions for Discussion:

1. Why do you think that those who work on the science of vision are particularly disposed to a localized rather than a distributive approach to brain processing?

2. To what extent is a strictly localized approach to brain processing ad odds with the fundamental principles of neuroplasticity?

VIII. Technological Applications

From medical rehabilitation to a joint operating system

HB: OK, now suppose I've been patiently reading all of this, but I'm not really interested in the particulars of neuroscience or competing theories. What I'm really interested in are the technological implications and applications.

"It's all very well and good that we can pick up monkeys' thoughts as they play video games or think about walking in order to move faraway robots so that they can get a treat, but how is this relevant to us? What sort of technologies can we develop from all of this to help people who have serious impairments, such as quadriplegics, those who have Parkinson's disease and others?

MN: I think that in the short run, over the next few years, the main impact of this thing that we call a "brain-machine interface", this paradigm that we've been talking about, will be in medical rehabilitation. There's no doubt about it.

Patients who are paralyzed will benefit from this possibility of bypassing the lesion and using brain activity to control a very wide variety of prosthetic devices: single limb, lower limb, whole body, upper limb and so forth. There is now a huge diversity of potential devices. Then there's communication. For people who cannot communicate, they will be able to use their brain activity to communicate. It's not only paralysis.

We are working here in the lab on prosthetic devices for Parkinson's disease that take advantage of the basic science that we've discussed regarding this new model of the brain. They would never work if the brain operated on the classic model.

Because of this view of the brain, we've created a prosthetic that uses the spinal cord, the surface of the spinal cord, to send an electrical signal to the brain to desynchronize the neurons that are firing together during Parkinson's, like an epileptic seizure,

HB: Oh, is that what Parkinson's really is?

MN: Well, it's a relatively new concept: that's what *I* believe it is. Parkinson's manifests itself, from a neurological point of view, as a seizure, as neurons firing rhythmically together. And too much order in the brain is pathological.

HB: So you're trying to decouple this?

MN: I decouple the neurons, I make them out of phase, by sending a desynchronizing signal from the spinal cord.

If the brain works as they say it does in textbooks, this should not work at all: it would have no effect whatsoever. In most animal models, in rats, mice, monkeys, and now in about 18 patients from around the world, the thing has proved to be very useful. There are other prosthetic devices to restore vision, sense of touch and so forth.

So that's the field of rehabilitation medicine—that's where you're going to see the impact of all of this very quickly.

I truly believe that there are going to be other ways of brain-machine interfaces, not only for patients with neurological disorders or psychiatric disorders. We will see this technology advance using non-invasive methods—no more sensors inside the brain but now from the outside. This will allow us to have a completely different experience of interacting with our computers and anything that is digitally controlled.

In a couple decades we will very likely be a part of our desktops, our laptops, our iPads or whatever: we will be a part of the operating system in the sense that we will be interacting directly with applications by thinking. And we'll get feedback. We are going to be assimilated into the operating system.

Questions for Discussion:

1. Might there be particular bioethical concerns associated with any of the medical technologies Miguel mentions?

2. How might brain-machine interfaces be used to "prove", or at least "demonstrate" consciousness? (Those particularly interested in this issue are referred Chapter 8 of the Ideas Roadshow conversation **The Limits of Consciousness** *with UCLA cognitive scientist Martin Monti)*

3. Do Miguel's speculations of how our brains will be "assimilated into the operating systems" of our computers fill you with excitement or with dread?

IX. Getting Metaphysical

The brain as reality-mediator

HB: Once again it's this notion of using what's around us concretely as a tool, looking at it from a brain-driven perspective, where the brain is actually imposing on the world. You wrote a book called *Beyond Boundaries*, and there's a clear sense both from that book and talking with you that you view the human body as a limit, almost as an arbitrary limit, which we are now poised to go beyond.

MN: Yes. That's what the definition of the sense of self used to be. The limit was *here*, at the last layer of the skin. I think the brain never cared about that. **We** cared. **We** had the feeling that this was the limit of myself, what distinguishes me from you. That's exactly the point.

Now I want to stretch this concept in a new book (*The True Creator of Everything: How the Human Brain Shaped the Universe as We Know It*) by saying that **everything** that surrounds us is created by the brain: the very notion of reality that we live with, this whole world—the political and economic system, our social relationships, our prejudices, our religion—*everything* that we claim to be tangible reality is this imposition from within our brain on the outside.

HB: OK, so now we can finally talk about this in a bit more detail, because you've mentioned this a couple times and I kept holding you back.

MN: Yes.

HB: I'm not sure what the actual claim is. I remember earlier when you talked about Carl Sagan and rewriting that "in the beginning" bit of *Cosmos* to focus on the brain instead of the cosmos.

Is the claim that there is no objective, independent reality outside of the human brain, and that everything we interpret as such is create by our brains?

Or is it that the only way we can get access to an objectively-existing external reality, the only way we can talk meaningfully about it, is by first recognizing that we necessarily have a fundamentally brain-centred perspective?

MN: It's the second way. It's the best way to reach this truth.

HB: Well, I suppose it's the only way.

MN: Yes, it's the only way. But there is no final truth. I mean, physicists like to say that our limit to understanding the universe is the speed of light because there's a sphere around us of time.

HB: Sure, there's a light barrier.

MN: Exactly. We cannot see beyond that barrier because light will never get to us. I'm putting up another barrier, which is the biology of our present brain that gives us one potential view of this reality. It's the best we can do. And science is probably the best we can do within that sphere.

HB: But the reality's there,

MN: There is a reality of course.

HB: Well, I'm with you. But since we're getting metaphysical, it's perhaps worth pointing out for completeness that there are people who think differently.

MN: I'm going to give the following example and then you'll see what I mean. If another brain would show up from a different civilization that evolved based on a different type of biology and created a central nervous system that is very different from ours, my contention would be that the explanation that they would provide for the universe would be different than ours. Even their scientists would have a different explanation.

HB: OK, but there's still a correct explanation out there, right? I mean, there's something out there.

MN: No, no, of course, of course. But that's the best we can do and I will explain why. When I look at the world, when I touch the world, there are no concepts out there. There's information. We create concepts. Our brain creates concepts. The concept of the proton, neutron, star, is a concept that a human brain created based on the evidence that he collected. So these concepts are not necessarily universal. It depends on the biology of our brains.

In fact some of the physicists may have touched on the limits of our biology like Gödel, his theorems are probably the limit of our logic. And that explains that our biology cannot definitively prove that something is true, even though it's true in a closed system.

HB: Mathematics, in fact, is perhaps even more interesting in that respect. Because one can talk about the objective reality of protons and neutrons, and how alien life might interpret that or might not interpret that.

MN: Exactly.

HB: But one can also talk about mathematical truth, and the Incompleteness Theorems and all the rest of that.

MN: For me, part of the beauty of this from an intellectual point of view is that it's finally possible to make this connection: that we might be able to talk to physicists and chemists and mathematicians about their world that they love through ours, which is the brain, and actually find a subject that is interesting for both sides.

Because there is no theory of everything. What I love about this is that when I read Stephen Hawking's first big book *A Brief History of Time*, he said on the first page, *"We are going to have the theory of everything in a few years."* His last book says it is impossible.

HB: He wasn't the only one who had that little development.

MN: Yes, I know, there are many. But in my lifetime as a professional scientist I have seen this. And from a narrow biological point of view I can actually relate to it. That's what I find intellectually very satisfying, very exciting: it's a good time to be alive doing neurobiology.

Questions for Discussion:

*1. How might the notion that "information is physical" relate to the concepts discussed in this chapter? (For more on the concept of the physicality of information, readers are referred to the Ideas Roadshow conversation **Cryptoreality** with University of Oxford and NUS physicist Artur Ekert)*

*2. Would an alien intelligence necessarily have an equivalent understanding of fundamental mathematical concepts, such as pi? (For more on this concept of so-called "Mathematical Platonism", readers are referred to the Ideas Roadshow conversation **Plato's Heaven: A User's Guide** with University of Toronto philosopher James Robert Brown)*

X. Final Questions

Big ones, dark ones and ontological ones

HB: Let me now turn to the potential dark side of all of this.

MN: Remember, I'm Brazilian. The glass is always half full.

HB: OK. But I'm not even suggesting that you should play this role. I'm imagining someone else reading this and saying, "*OK, this Nicolelis guy is able to make all sorts of wonderful machines that are able to help people who have difficulties because we are now able to process mental information and read brain waves and so forth. But what about the prospect of going the other way, what you sometimes refer to as "brain-to-brain" information transfer, where we can imagine someone or something imposing signals on our brains through technology?*"

So if I'm a science-fiction-oriented individual, and I'm worried about things—because maybe I've watched too many science fiction movies or whatever—I might be thinking, "*My gosh, what are the negative things that could come out of this? Someone might actually be able to have some sort of a brain ray that could force me to think different thoughts, or they could read my thoughts, or—*"

MN: It's already happening. It's called TV.

HB: OK. Fair enough. But in *Beyond Boundaries* you mentioned something scarily, eerily similar to this. There's this guy Delgado with a radio-controlled device and a bull comes roaring at him and he's able to press this button and the bull becomes pacified. Might it be possible for us to develop something like that for humans as well?

I mean you're talking about all these wonderful things: you're developing ways to help people, you're developing brain-to-brain communication in rats—

MN: Do you know the motivation of the brain-to-brain experiment? Do you know why we did it? The motivation was to see if we could make the second brain's representation of the sensory cortex, tactile cortex, assimilate the whiskers of the first animal. So we will now have an animal that has a representation of its own body plus the body of its companion.

HB: So again this "beyond boundaries" notion.

MN: Yes. Rats don't have a sense of self that we can speak of—

HB: Well, who knows?

MN: Right: who knows? Exactly. There's a debate, but I don't want to get into that.

But in any event we show such an avid plasticity of an assimilator, that for over a month it can actually create a representation in one animal of the tactile sensor array of another animal. That's the most important result and almost nobody paid attention to it.

Sure, we wanted to show the concept; and I have to point out that science fiction has already done it. If you read Arthur C. Clarke's *3001: The Final Odyssey*, the whole topic is a civilization—us, a thousand years from now— that developed "braincaps" that allowed them to communicate and acquire knowledge.

So there's nothing that we have done that's even close to that idea—allowing you to learn Latin just by receiving signals in your brain, there's no way that anyone could do that right now.

Even Delgado's scary scenario with the bull—for which he paid a pretty heavy price, by the way, he was kicked out of Yale and had to leave the United States for some time—that is nothing: that is just to stop a movement. Nobody that I know can use any technology remotely linked to this to control anybody's behaviour.

HB: OK, so you wouldn't worry about this being the thin edge of the wedge or anything like that?

MN: Not really, no. Sure you can always think that there must be some crazy lunatic who's thinking about how to take advantage of this in a bad way, but that's the same for everything.

HB: But that's not science right? Science is simply about developing understanding and learning.

MN: Yes. Our job is to develop things, ideas, notions. It's society's job to regulate what will be used and what will not. I cannot decide. I have my own opinions, but how people are going to use this technology is another matter.

As I said, I see things from the bright side. I see that one day one of us, or one of our grandchildren, will have a stroke; and instead of remaining aphasic for the rest of his or her days, we'll have a bypass that will communicate the right hemisphere to the left and that person will be able to rehabilitate the damage to the Broca area and all of the cortical areas of the left area due to the massive stroke, and they will speak again. Or be able to communicate. That's what I'm aiming at.

But of course any piece of technology can be used for terrible outcomes. Look at what so often happens to soccer balls in this country. The soccer ball should be delicately caressed and passed, and look at the tragedy that occurs to those beautiful objects every day.

HB: Right—I had forgotten about all of that. It's good to know that, as an irrepressibly optimistic football-obsessed fellow you so heartily represent your country's core values.

One last question: You're obviously a very passionate individual. You've dedicated your life to science. You've clearly done great things. If you were in a situation where you were face to face with some omniscient being, what would you ask him? What are the key things that you are dying to know the answers to?

MN: Well, some of the ones that we've touched on here: what is the number? How many neurons?

HB: How many makes a fundamental thinking unit?

MN: Yes. What is the minimum number? Why this way? How does it feel to use an avatar to touch a virtual surface? What do you feel? Our monkeys cannot tell us, but they are literally using a virtual arm to touch a virtual object and they can distinguish them based on the virtual texture of that surface.

These would be the questions. And for me the ultimate request would be, "*Please give us a little more time to try to answer it*". I don't just want the answers: I want the pathway to find them. I think science is about that.

I don't worry about the fact that we may never touch the truth because invariably the truth will never be found anyway. What I want is the opportunity to pursue it. That, I think, is what the right scientific mission is: this journey towards this elusive truth. That's what matters.

HB: That's a great note to end on. Is there anything else we should mention, you think?

MN: No, I think we talked about pretty much everything.

HB: Well, it was a pleasure Miguel. Thank you very much for your time.

MN: Likewise. Thank you.

Questions for Discussion:

*1. If there **was** an opportunity to just learn something by receiving signals to your brain, would you do it?*

2. To what extent do you agree or disagree with Miguel's statement "The truth will never be found anyway"?

3. How has this conversation changed the way you view the brain and possible new technologies associated with it?

Continuing the Conversation

Readers interested in knowing more about Miguel's ideas are encouraged to read his books: *Beyond Boundaries: The New Neuroscience of Connecting Brains With Machines—and How It Will Change Our Lives*, *The True Creator of Everything: How the Human Brain Shaped the Universe as We Know It* and *The Relativistic Brain: How It Works and Why It Cannot be Simulated by a Turning Machine* (with Ronald Cicurel).

Ideas Roadshow Collections

Each Ideas Roadshow collection offers 5 separate expert conversations presented in an accessible and engaging format.

- *Conversations About Anthropology & Sociology*
- *Conversations About Astrophysics & Cosmology*
- *Conversations About Biology*
- *Conversations About History, Volume 1*
- *Conversations About History, Volume 2*
- *Conversations About History, Volume 3*
- *Conversations About Language & Culture*
- *Conversations About Law*
- *Conversations About Neuroscience*
- *Conversations About Philosophy, Volume 1*
- *Conversations About Philosophy, Volume 2*
- *Conversations About Physics, Volume 1*
- *Conversations About Physics, Volume 2*
- *Conversations About Politics*
- *Conversations About Psychology, Volume 1*
- *Conversations About Psychology, Volume 2*
- *Conversations About Religion*
- *Conversations About Social Psychology*
- *Conversations About The Environment*
- *Conversations About The History of Ideas*

All collections are available as both eBook and paperback.

www.ingramcontent.com/pod-product-compliance
Lightning Source LLC
Chambersburg PA
CBHW020246030426
42336CB00010B/645